New Developments in Soliton Research

NEW DEVELOPMENTS IN SOLITON RESEARCH

L.V. CHEN
EDITOR

Nova Science Publishers, Inc.
New York

For permission to use material from this book please contact us:
Telephone 631-231-7269; Fax 631-231-8175
Web Site: http://www.novapublishers.com

LIBRARY OF CONGRESS CATALOGING-IN-PUBLICATION DATA
New developments in soliton research / L.V. Chen (editor).
p. cm.
Includes bibliographical references and index.
ISBN 1-59454-561-8
1. Solitons--Research. I. Chen, L. V.
QC174.26.W28N49 2006
531'.1133--dc22 2005027915

Published by Nova Science Publishers, Inc. ✣ New York

CONTENTS

PREFACE

Since their discovery a mere thirty years ago, solitons have been invoked to explain such diverse phenomena as: The long lived 'giant red spot' in the highly turbulent Jovian atmosphere. The famous Fermi-Pasta-Ulam paradox wherein a nonlinearly coupled lattice of particles does not display the "expected" equipartition of energy among available modes. Ion-acoustic waves in a plasma. Energy storage and transfer in proteins via the Davydov soliton. The propagation of short laser pulses in optical fibers over long distances with negligible shape change. This new volume presents important research from around the globe.

In Chapter 1 a series of solutions with higher baryon numbers in the chiral quark soliton model are reported. *The chiral quark soliton model* (CQSM) is a simple quark model that incorporates the basic features of QCD, e.g. the chiral symmetry and its breakdown accompanied by the appearance of the Goldstone bosons. It was shown that the baryon number one (B = 1) solution provides correct observable as a nucleon including mass, electromagnetic value, spin carried by quarks, parton distributions and octet, decuplet SU(3) baryon spectra. The B = 2 axially symmetric soliton solution was obtained numerically by one of us (N. Sawado). For B $\geq$ 3, in a series of our papers we obtained B = 3 ⊔ 9, 17 minimal energy soliton solutions with point-like symmetries and B = 3 ⊔ 5 axially symmetric saddle-point solutions, using rational map ansatz constructed for multi-baryon number skyrmions. An interesting property of the solutions is that the symmetry of the background configuration is reflected in the degeneracy of the valence quark spectra. For instance, the resultant quark bound spectra are doubly degenerate for the axisymmetric solitons and are triply degenerate for B = 3 tetrahedrally symmetric solitons. These results confirm the existence of the quark shell structure. The shells consist of the four-fold degenerate ground state and higher levels with various patterns, which are realized by the interplay of two symmetries, $SU(2)_L \times SU(2)_R$ symmetry for the quarks and the symmetries of the chiral fields. To obtain correct physical observable, the quantum corrections are necessary. We shall show the quantum states of the axisymmetric solitons within the collective quantization. Upon quantization, various observable spectra of the chiral solitons are obtained. According to the Finkelstein-Rubinstein constraints, the quantum numbers of the solitons coincide with the physical observations only for B = 2 and 4 while B = 3 and 5 do not. The SU(3) extension of the B = 2 soliton is also studied to predict various strange dibaryon states within this model.

We review in Chapter 2 our recent construction of soliton and topological solutions of noncommutative gauge theory on a fuzzy sphere. To reach these goals we have employed several methods, i.e. the matrix model approach and the projective module one, introducing a new general link between them. In the case of search for solitons we have firstly introduced an ansatz into the general set of classical equations of motion of the matrix model, obtaining a general class of solutions. Between them we have selected those admitting a classical limit, which in turn require that a certain coupling constant _ of the model behaves as $\lambda = \frac{1}{2} + \frac{\lambda_0}{N}$ in the N → ∞ limit. In this way we find classical solitons characterized by the angular momentum l = 2. More general type of ansatze are required to search for solitons with general angular momentum l. This technique is not enough to find topological solutions, whose classical limit depends on all values of the angular momentum. We appeal to another method, based on projective modules, to find noncommutative monopoles inside the matrix model, tending to classical monopoles of the sphere. The projective modules are built making use of the the principal Hopf fibration: $S^3 \rightarrow S^2$, and its extension to the fuzzy sphere case. The link with the matrix models is possible only if the projectors can be recast in the form of a ket-bra combination. Once that the resulting vector valued operator is known, it is possible to construct topological solutions of the matrix model at the noncommutative level.

Nonlinear interaction of an intense electromagnetic (EM) beam with relativistically hot electron-positron plasma is investigated in Chapter 3 by invoking the variational principle and numerical simulation, resting on the model of generalized nonlinear Schr¨odinger equation with saturating nonlinearity. The present analysis shows the dynamical properties including the possibilities of trapping and wave-breaking of EM beams. These properties of EM beams may give a significant clue for the gamma-ray burst.

The reaction-diffusion as well as Klein-Gordon differential-difference equation (DDE) models are revisited in Chapter 4, and categorized as generic [in the presence of a Peierls-Nabarro (PN) barrier for their static solutions] and non-generic (in the absence of one such). The various available ways to construct models without PN barriers are detailed and their respective features are highlighted. This categorization of DDE's is used to identify the properties of travelling waves that are present in these models upon the action of an external force. The onset of the relevant, so-called depinning transition (to travelling waves) in both the generic and the non-generic discretization case is analyzed and its critical properties are elucidated using analytical calculations (normal form, energy landscape and formal expansion considerations), as well as numerical computations.

We consider the Lagrangian description of the soliton sector of the so-called affine $\hat{sl}(3)$ Toda model coupled to matter (Dirac) fields (ATM). The theory is treated as a constrained system in the contexts of the Faddeev-Jackiw, the symplectic, as well as the master Lagrangian approaches. We exhibit the master Lagrangian nature of the model from which generalizations of the sine-Gordon (GSG) or the massive Thirring (GMT) models are derivable. The GMT model describes $N_f = 3$ [number of positive roots of *su*(3)] massive Dirac fermion species with current-current interactions amongst all the U(1) species currents; on the other hand, the GSG theory corresponds to $N_b = 2$ [rank of the *su*(3) Lie algebra] independent Toda fields (bosons) with a potential given by the sum of three SG cosine terms. The dual description of the model is further emphasized by providing some on shell relationships between bilinears of the GMT spinors and the relevant expressions of the GSG fields. In this way, in the **first part** of Chapter 5, we exhibit the strong/weak coupling phases and the (generalized) soliton/particle correspondences of the model at the classical level. In the **second part** of the chapter we give a full Lie algebraic formulation of the duality at the level of the equations of motion written in matrix form. The effective off-critical $\hat{sl}$ (3) ATM action is written in terms of the Wess-Zumino-Novikov-Witten (WZNW) action plus some kinetic terms for the spinors and scalar-spinor interaction terms. Moreover, this theory still presents a remarkable equivalence between the Noether and topological currents, describes the soliton sector of the original model and turns out to be the master Lagrangian describing the GMT and GSG models.

A dynamical bifurcation mechanism related to changes of metric signature is used in Chapter 6 to explain the formation of topological defects experimentally observed and defined in 1986 as Falaco Solitons. The Falaco Solitons are topologically coherent structures created experimentally by a macroscopic rotational dynamics in a continuous media with a discontinuity surface, such as that found in a swimming pool. The topological structure of Falaco Solitons replicates certain features found at all physical scales, from spiral arm galaxies and cosmic strings to submicroscopic hadrons. The easy to replicate experiment indicates that the creation of "stationary" thermodynamic states (or solitons) far from equilibrium can be globally stabilized. The Falaco solitons represent a paradigm for explaining a spin pairing mechanism in the microscopic Fermi surface, the development of dimpled vortex structures in rotating Bose-Einstein Condensates, the confinement problem of sub-microscopic quarks on the end of a string connecting branes, and the quantized needle radiation of as photon in terms of a string connecting two Falaco Soliton dimples located on concentric light cone shells.

In: New Developments in Soliton Research
Editor: L.V. Chen pp. 1-52

ISBN 1-59554-561-8

Chapter 1

MULTI-SOLITON SOLUTIONS IN THE CHIRAL QUARK SOLITON MODEL

Nobuyuki Sawado and Noriko Shiiki
Department of Physics,
Tokyo University of Science,
Noda, Chiba 278-8510, Japan

Abstract

In this article a series of solutions with higher baryon numbers in the chiral quark soliton model are reported. *The chiral quark soliton model* (CQSM) is a simple quark model that incorporates the basic features of QCD, e.g. the chiral symmetry and its breakdown accompanied by the appearance of the Goldstone bosons. It was shown that the baryon number one ($B = 1$) solution provides correct observable as a nucleon including mass, electromagnetic value, spin carried by quarks, parton distributions and octet, decuplet $SU(3)$ baryon spectra. The $B = 2$ axially symmetric soliton solution was obtained numerically by one of us (N. Sawado). For $B \geq 3$, in a series of our papers we obtained $B = 3 \sim 9, 17$ minimal energy soliton solutions with point-like symmetries and $B = 3 \sim 5$ axially symmetric saddle-point solutions, using rational map ansatz constructed for multi-baryon number skyrmions. An interesting property of the solutions is that the symmetry of the background configuration is reflected in the degeneracy of the valence quark spectra. For instance, the resultant quark bound spectra are doubly degenerate for the axisymmetric solitons and are triply degenerate for $B = 3$ tetrahedrally symmetric solitons. These results confirm the existence of the quark shell structure. The shells consist of the four-fold degenerate ground state and higher levels with various patterns, which are realized by the interplay of two symmetries, $SU(2)_L \times SU(2)_R$ symmetry for the quarks and the symmetries of the chiral fields. To obtain correct physical observable, the quantum corrections are necessary. We shall show the quantum states of the axisymmetric solitons within the collective quantization. Upon quantization, various observable spectra of the chiral solitons are obtained. According to the Finkelstein-Rubinstein constraints, the quantum numbers of the solitons coincide with the physical observations only for $B = 2$ and 4 while $B = 3$ and 5 do not. The $SU(3)$ extension of the $B = 2$ soliton is also studied to predict various strange dibaryon states within this model.

1 Introduction

Although QCD is generally accepted as the underlying theory of the strong interaction, most low- and medium-energy nuclear phenomenology may be successfully described in terms of the hadronic degrees of freedom. Investigations for deuteron photodisintegration and deep inelastic scattering of leptons by nuclei suggest the necessity of including quark degrees of freedom [1, 2, 3, 4, 5]. For this reason it was suggested that nuclear theory should be reformulated to take into account the underlying quark theory. However, QCD is too hard to get insight into the low- and medium-energy nuclear phenomenology since the coupling constant become large at these energy scales and one can not perform perturbation as in the other gauge theories. It is therefore necessary to formulate low-energy effective theories for the strong interaction.

The chiral quark soliton model is one of such QCD effective theories including quark degrees of freedom and baryons as a chiral soliton. In 1997 Diakonov *et al* predicted an exotic state with strangeness $+1$ within the topological soliton picture of baryons [6], and remarkably such resonance state has been experimentally discovered recently [7]. In a naive point of view, this new exotic baryon can be interpreted as a five quark bound state, opening up a new paradigm of the nuclei as a multi-quark bound state.

The chiral quark soliton model (CQSM) was developed in 1980's as a low-energy effective theory of QCD. Since it includes the Dirac sea quark contribution and explicit valence quark degrees of freedom, the model interpolates between the constituent quark model and the Skyrme model [8, 9, 10, 11, 12]. The CQSM is derived from the instanton liquid model of QCD vacuum and incorporates the nonperturbative feature of the low-energy QCD, spontaneous chiral symmetry breaking. It has been shown that the $B = 1$ solution provides correct observable as a nucleon including mass, electromagnetic value, spin carried by quarks, parton distributions and octet $SU(3)$ baryon spectra.

For $B = 2$, the stable axially symmetric soliton solution was found in Eq. [13]. The solution exhibits doubly degenerate bound spectra of the quark orbits in the background of the axially symmetric chiral field with winding number two. Upon quantization, various dibaryon spectra were obtained, showing that the quantum numbers of the ground state coincides with those of a physical deuteron [14, 15]. For $B \geq 3$, the Skyrme model shows that the soliton solutions have discrete, crystal-like symmetries [16, 17]. From the similarity of the chiral field action between the Skyrme model and the CQSM one can expect that soliton solutions in the CQSM have same symmetries as skyrmions with the same baryon number. Since it is too complicated to perform a numerical computation if one imposes such discrete symmetries directly on the chiral fields, Houghton, Manton and Sutcliffe thus proposed remarkable ansatz, rational map ansatz, for multi-skyrmions [18]. Applying this ansatz to the CQSM we obtained multi-chiral quark soliton solutions with point-like symmetries for $B = 3 \sim 9, 17$ as well as saddle-point solutions for $B = 5, 9$ [19, 20]. The solutions exhibit a large degeneracy and mass gap in the valence quark orbits. These results confirm the existence of the quark shell structure. Esepecially the large degeneracy implies that our solutions with such polyhedral symmetries may be the lowest-lying configuration.

The solitons that we obtain are classical objects and therefore must be quantized to assign definite spin and isospin to them. Quantization of the solitons can be performed semiclassically for their rotational zero modes. Quantizing the solutions with discrete symmetries is, however, a formidable task in CQSM. Thus, before embarking those discrete symmetries, it will be instructive to study axially symmetric solutions which are much simpler [21]. Besides, considering the fact that for some higher baryon numbers, the ground states of the skyrmions do not agree with the experimental observation [22] , the possibility that axially symmetric solutions may provide correct ground states can not be excluded. In fact it was found in Ref. [23] that the axially symmmetric BPS monopoles up to chrage five have lower energies than those of discrete symmetries. We therefore investigate classical and quantum multi-soliton solutions in the CQSM with axial symmetry up to $B = 5$.

Since the first prediction of the H-particle in a MIT bag model calculation [24], there have been many efforts to study the spectrum of dibaryonic systems including strangeness. We shall apply our formulation to study these six quark states constituting a dibaryon and make a prediction for its mass spectra.

In Sec. 2 the formulation of the chiral quark soliton model is introduced. In Sec. 3 we obtain axially symmetric soliton solutions for $B = 2 \sim 5$ and solutions with the polyhedral symmetries for $B = 3 \sim 9$ including saddle-point configurations. The relation between the symmetries of the chiral fields and the degeneracy of the valence quark spectra is discussed in Sec. 4 from a group theoretical point of view. In Sec. 5 we perform zero mode quantization for the obtained classical solitons. Imposing the Finkelstein-Rubinstein constraints on the states, the ground states of the axially symmetric solitons are constructed and examined if they agree with the experimental observation. Conclusions and discussions are in Sec.6.

2 The Chiral Quark Soliton Model : General Formalism

The CQSM is derived from the instanton liquid model of the QCD vacuum and incorporates the nonperturbative feature of the low-energy QCD, spontaneous chiral symmetry breaking. The vacuum functional is defined by Ref. [8]

$$\mathcal{Z} = \int \mathcal{D}\pi \mathcal{D}\psi \mathcal{D}\psi^\dagger \exp\left[i \int d^4x\, \bar{\psi}\,(i\partial\!\!\!/ - MU^{\gamma_5})\,\psi\right] \tag{1}$$

where the SU(2) matrix

$$U^{\gamma_5} = \frac{1+\gamma_5}{2}U + \frac{1-\gamma_5}{2}U^\dagger \quad \text{with} \quad U = \exp\left(i\boldsymbol{\tau}\cdot\boldsymbol{\pi}/f_\pi\right)$$

describes chiral fields, ψ is quark fields and M is the constituent quark mass. f_π is the pion decay constant and experimentally $f_\pi \sim 93$MeV.

The $B = 1$ soliton solution has been studied in detail at classical and quantum level in Refs. [8, 9, 10, 11, 12]. To obtain solutions with $B > 1$, we shall employ the chiral fields with winding number B in the Skyrme model as the background of quarks, which can be justified as follows.

In Eq. (1), performing the functional integral over ψ and $\psi^\dagger$ fields, one obtains the effective action

$$S_{\text{eff}}(U) = -iN_c\text{Sp}\ \ln iD = -iN_c \log\det iD, \tag{2}$$

where $iD = i\partial\!\!\!/ - MU^{\gamma_5}$ is the Dirac operator. The classical solutions can be obtained by the extremum condition of (2) with respect to U. For this purpose, let us consider the derivative expansion of the action [25, 26, 12]. Up to quartic terms, we have,

$$S_{\text{eff}}(U) = \int d^4x\Big[-C\text{tr}(L_\mu L^\mu) + \frac{N_c}{32\pi^2}\text{tr}\Big\{\frac{1}{12}[L_\mu, L_\nu]^2 - \frac{1}{3}(\partial_\mu L^\mu)^2 + \frac{1}{6}(L_\mu L^\mu)^2\Big\}\Big], \quad (3)$$

where $L_\mu = U^\dagger \partial_\mu U$. The baryon number B can be calculated in terms of the topological charge,

$$B = -\frac{1}{24\pi^2}\epsilon_{ijk}\int d^3x\text{tr}(L_i L_j L_k)\,. \quad (4)$$

Suitably adjusting the coefficients C, one can identify the first two terms of Eq. (3) with the Skyrme model action. However, the 4th order terms tend to destabilize solutions and no stable classical solution can be obtained from the above action [25, 27]. Nevertheless, because of their similarity, it will be justified to adopt the configurations of the solutions in the Skyrme model to chiral fields in the CQSM.

In the CQSM, the number of valence quark is associated with the baryon number such that the baryon number B soliton consist of $N_c \times B$ valence quarks. If the quarks are strongly bound inside the soliton, their binding energy become large and the valence quarks can not be observed as positive energy particles [28, 29]. Thus, one gets the picture of the topological soliton model in the sense that the baryon number coincide with the winding number of the background chiral field when the valence quarks occupy all the levels diving into negative energy region.

Let us rewrite the effective action in Eq. (2) as

$$S_{\text{eff}} = -iN_c \log\det(i\partial\!\!\!/ - MU^{\gamma_5}) = -iN_c \log\det\big(i\partial_t - H(U^{\gamma_5})\big) \quad (5)$$

where

$$H(U^{\gamma_5}) = -i\alpha\cdot\nabla + \beta MU^{\gamma_5}\,. \quad (6)$$

The classical energy of the soliton can be estimated from the quark determinant in Eq. (5) [30, 31]. We introduce the eigenstates of operators, $i\partial_t - H(U^{\gamma_5})$ and $H(U^{\gamma_5})$, such that

$$H(U^{\gamma_5})\phi_\mu(\boldsymbol{x}) = E_\mu\phi_\mu(\boldsymbol{x})\,, \quad (7)$$

$$\big(i\partial_t - H(U^{\gamma_5})\big)\Psi_{\mu,n} = \lambda_{\mu,n}\Psi_{\mu,n}\,, \quad (8)$$

where $\Psi_{\mu,n} = e^{-i\omega_n t}\phi_\mu$ and $\lambda_{\mu,n} = -E_\mu + \omega_n$. Imposing on $\Psi_{\mu,n}$ the anti-periodicity condition, $\Psi_{\mu,n}(\boldsymbol{x}, T) = -\Psi_{\mu,n}(\boldsymbol{x}, 0)$, reads

$$\omega_n T = (2n+1)\pi. \quad (9)$$

The determinant in Eq. (5) then becomes

$$\begin{aligned}\det(i\partial_t - H) &= \prod_{\mu,n}\lambda_{\mu,n} = \prod_{\mu,n}\Big(-E_\mu + \frac{(2n+1)\pi}{T}\Big) \\ &= C\prod_{\mu,n\geq 0}\Big(1 - \frac{|E_\mu|^2T^2}{(2n+1)^2\pi^2}\Big) = C\prod_\mu \cos\Big(\frac{1}{2}|E_\mu|T\Big) \\ &= \frac{C}{2}\exp\Big(i\frac{1}{2}\sum_\mu|E_\mu|T\Big)\prod_\mu\Big(1+\exp(-i|E_\mu|T)\Big) \quad (10)\end{aligned}$$

where

$$C = \prod_{n\geq 0}\Big(-\frac{(2n+1)^2\pi^2}{T^2}\Big)$$

and the product formula for the cosine function $\cos(z) = \prod_{n\geq 1}^{\infty}(1-4z^2/(2n-1)^2\pi^2)$ has been used. Inserting (10) into Eq. (5), one obtains

$$S_{\text{eff}} = -N_c T\sum_{\mu} n_\mu |E_\mu| + N_c T\frac{1}{2}\sum_{\mu} |E_\mu|\,, \tag{11}$$

where n_μ is the valence quark occupation number which takes values only 0 or 1. Correspondingly, the classical energy is given by

$$E_{\text{static}} = E_{\text{val}} + E_{\text{vac}} \tag{12}$$

where

$$E_{\text{val}} = N_c\sum_{\mu} n_\mu |E_\mu|\,,\ E_{\text{vac}} = -\frac{1}{2}N_c\sum_{\mu}|E_\mu|\,,$$

representing the valence quark and sea quark contribution to the total energy respectively.

The effective action $S_{\text{eff}}(U)$ is ultraviolet divergent and hence must be regularized. Using the proper-time regularization scheme [32], we can write

$$\begin{aligned} S_{\text{eff}}^{\text{reg}}[U] &= \frac{i}{2}N_c\int_{1/\Lambda^2}^{\infty}\frac{d\tau}{\tau}\text{Sp}\left(\text{e}^{-D^\dagger D\tau} - \text{e}^{-D_0^\dagger D_0\tau}\right) \\ &= \frac{i}{2}N_c T\int_{-\infty}^{\infty}\frac{d\omega}{2\pi}\int_{1/\Lambda^2}^{\infty}\frac{d\tau}{\tau}\text{Sp}\left[\text{e}^{-\tau(H^2+\omega^2)} - \text{e}^{-\tau(H_0^2+\omega^2)}\right] \end{aligned} \tag{13}$$

where D_0 and H_0 are operators with $U=1$. The total energy is then given by

$$E_{\text{static}}[U] = E_{\text{val}}[U] + E_{\text{vac}}[U] - E_{\text{vac}}[U=1] \tag{14}$$

where

$$E_{\text{val}} = N_c\sum_i E_{\text{val}}^{(i)},\ E_{\text{vac}} = N_c\sum_{\mu}\left\{\mathcal{N}(E_\mu)|E_\mu| + \frac{\Lambda}{\sqrt{4\pi}}\exp\left[-\left(\frac{E_\mu}{\Lambda}\right)^2\right]\right\}$$

with

$$\mathcal{N}(E_\mu) = -\frac{1}{\sqrt{4\pi}}\Gamma\left(\frac{1}{2},\left(\frac{E_\mu}{\Lambda}\right)^2\right)$$

and $E_{\text{val}}^{(i)}$ is the valence energy of the i th valence quark. Λ is a cutoff parameter evaluated by the condition that the derivative expansion of Eq. (13) reproduces the pion kinetic term with the correct coefficient, *i.e.*,

$$f_\pi^2 = \frac{N_c M^2}{4\pi^2}\int_{1/\Lambda^2}^{\infty}\frac{d\tau}{\tau}\text{e}^{-\tau M^2}\,. \tag{15}$$

In this model, the constituent quark mass M is the only free parameter and we take the value $M = 400$ MeV, in which the observable of the nucleon and the delta are well reproduced [11]. From Eq. (15) and by using the values of M, f_π, we obtain $\Lambda \sim 637$ MeV.

For $B = 1$, one imposes a spherically symmetric ansatz (*hedgehog ansatz*)

$$U(\boldsymbol{r}) = \exp(iF(r)\hat{\boldsymbol{r}}\cdot\boldsymbol{\tau}) = \cos F(r) + i\hat{\boldsymbol{r}}\cdot\boldsymbol{\tau}\sin F(r)\,, \tag{16}$$

with the boundary condition for the profile function $F(r)$

$$F(0) = -\pi, \quad F(\infty) = 0\,. \tag{17}$$

Substituting the ansatz (16) into Eq. (4) under the boundary condition (17), one gets

$$B = \frac{1}{2\pi^2}\int_0^\infty \frac{\sin^2 F(r)}{r^2}\frac{dF(r)}{dr}4\pi r^2 dr = \frac{1}{\pi}\left[F + \frac{\sin 2F}{2}\right]_{-\pi}^{0} = 1\,. \tag{18}$$

The one-quark hamiltonian (6) becomes

$$H(U^{\gamma_5}) = -i\alpha\cdot\nabla + \beta M(\cos F(r) + i\gamma_5\hat{\boldsymbol{r}}\cdot\boldsymbol{\tau}\sin F(r))\,. \tag{19}$$

This hamiltonian does not commute with the total angular momentum $\boldsymbol{J}$ nor the isospin $\boldsymbol{\tau}/2$ but commute with the grand spin operator $\boldsymbol{K} = \boldsymbol{J} + \boldsymbol{\tau}/2$. H also commutes with the parity operator $\mathcal{P} = \gamma_0$. Hence the one-quark eigenstates are labeled by the $K = 0, 1, 2, \cdots$ and the parity $\mathcal{P} = \pm$. The three valence quarks occupy their lowest states $K^{\mathcal{P}} = 0^+$ and responsible for the baryon number of the soliton $(= 1)$. In this context, the baryon number is not a topological origin unlike the Skyrme @model.

Field equations for the chiral fields can be obtained by demanding that the total energy in Eq. (14) be stationary with respect to variation of the profile function $F(r)$,

$$\frac{\delta}{\delta F(r)}E_{\text{static}} = 0\ ,$$

which produces

$$S(r)\sin F(r) = P(r)\cos F(r), \tag{20}$$

where

$$S(r) = N_c\sum_\mu (n_\mu\theta(E_\mu) + \text{sign}(E_\mu)\mathcal{N}(E_\mu))\langle\mu|\gamma^0\delta(|x| - r)|\mu\rangle\,, \tag{21}$$

$$P(r) = N_c\sum_\mu (n_\mu\theta(E_\mu) + \text{sign}(E_\mu)\mathcal{N}(E_\mu))\langle\mu|i\gamma^0\gamma^5\hat{\boldsymbol{r}}\cdot\boldsymbol{\tau}\delta(|x| - r)|\mu\rangle\,. \tag{22}$$

The procedure to obtain self-consistent solutions of Eq. (20) is that 1) solve the eigenequation in the hamiltonian (19) under an assumed initial profile function $F_0(r)$, 2) use the resultant eigenfunctions and eigenvalues to calculate $S(r)$ and $P(r)$, 3) solve (20) to obtain a new profile function, 4) repeat 1) – 3) until the self-consistency is attained.

The calculated energy of the $B = 1$ soliton is $E_{\text{static}} = 1192$ MeV with the constituent quark mass $M = 400$ MeV.

3 The Classical Configurations

3.1 The Axially Symmetric Configuration

It is known that the minimal energy solution for $B > 1$ is not spherically symmetric. In the Skyrme model the configuration with $B = 2$ is axially symmetric [33, 34, 35, 36] and can be written by

$$U(\boldsymbol{x}) = \cos F(\rho, z) + i\boldsymbol{\tau} \cdot \hat{\boldsymbol{n}} \sin F(\rho, z), \tag{23}$$

where

$$\hat{\boldsymbol{n}} = (\sin\Theta(\rho, z) \cos m_{\rm w}\varphi, \sin\Theta(\rho, z) \sin m_{\rm w}\varphi, \cos\Theta(\rho, z)) \tag{24}$$

and $m_{\rm w}$ is the winding number of the pion fields. We shall use this configuration in the background to obtain axially symmetric chiral quark solitons.

The extremum conditions for the total energy

$$\frac{\delta}{\delta F(\rho, z)} E_{\rm static}[U] = 0\,, \quad \frac{\delta}{\delta\Theta(\rho, z)} E_{\rm static}[U] = 0 \tag{25}$$

yield the following equations of motion for the profile functions,

$$R^T(\rho, z) \cos\Theta(\rho, z) = R^L(\rho, z) \sin\Theta(\rho, z)\,, \tag{26}$$

$$S(\rho, z) \sin F(\rho, z) = P(\rho, z) \cos F(\rho, z) \tag{27}$$

where

$$P(\rho, z) = R^T(\rho, z) \sin\Theta(\rho, z) + R^L(\rho, z) \cos\Theta(\rho, z)\,. \tag{28}$$

In terms of eigenfunction ϕ in Eq. (7), R^T,R^L and S are given by

$$R^T(\rho, z) = R^T_{\rm val}(\rho, z) + R^T_{\rm vac}(\rho, z)\,, \tag{29}$$

$$R^L(\rho, z) = R^L_{\rm val}(\rho, z) + R^L_{\rm vac}(\rho, z)\,, \tag{30}$$

$$S(\rho, z) = S_{\rm val}(\rho, z) + S_{\rm vac}(\rho, z) \tag{31}$$

where

$$R^T_{\rm val}(\rho, z) = \sum_i \int d\varphi \bar{\phi}_i(\rho, \varphi, z) i\gamma_5(\tau_1 \cos m_{\rm w}\varphi + \tau_2 \sin m_{\rm w}\varphi)\phi_i(\rho, \varphi, z)\,,$$

$$R^T_{\rm vac}(\rho, z) = \sum_\mu \mathcal{N}(E_\mu)\,{\rm sgn}(E_\mu) \int d\varphi \bar{\phi}_\mu(\rho, \varphi, z) i\gamma_5 \times(\tau_1 \cos m_{\rm w}\varphi + \tau_2 \sin m_{\rm w}\varphi)\phi_\mu(\rho, \varphi, z)\,,$$

$$R^L_{\rm val}(\rho, z) = \sum_i \int d\varphi \bar{\phi}_i(\rho, \varphi, z) i\gamma_5\tau_3\phi_i(\rho, \varphi, z)\,,$$

$$R^L_{\rm vac}(\rho, z) = \sum_\mu \mathcal{N}(E_\mu)\,{\rm sgn}(E_\mu) \int d\varphi \bar{\phi}_\mu(\rho, \varphi, z) i\gamma_5\tau_3\phi_\mu(\rho, \varphi, z)\,,$$

$$S_{\rm val}(\rho, z) = \sum_i \int d\varphi \bar{\phi}_i(\rho, \varphi, z)\phi_i(\rho, \varphi, z)\,,$$

$$S_{\rm vac}(\rho, z) = \sum_\mu \mathcal{N}(E_\mu)\,{\rm sgn}(E_\mu) \int d\varphi \bar{\phi}_\mu(\rho, \varphi, z)\phi_\mu(\rho, \varphi, z)\,,$$

and subscripts, vac and val, represent the vacuum and valence quark contributions respectively. The boundary conditions for the profile functions were constructed by Braaten and Carson [37];

$$F(\rho, z) \to 0 \ \text{as} \ \rho^2 + z^2 \to \infty,$$
$$F(0,0) = -\pi, \quad \Theta(0,z) = \begin{cases} 0, & z > 0 \\ \pi, & z < 0 \end{cases} . \tag{32}$$

In Fig. 1, we show the spectrum of the quark orbits in the background of chiral fields with winding number $m_{\rm w} = 2, 4$, as a function of the size parameter X. The axially symmetric profile functions are parameterized by X as

$$\begin{aligned} F(\rho, z) &= -\pi + \pi\sqrt{\rho^2 + z^2}/X \ \text{for} \ \sqrt{\rho^2+z^2} \le X \\ &= 0 \qquad\qquad\qquad\qquad \text{otherwise}, \end{aligned} \tag{33}$$
$$\Theta(\rho, z) = \tan^{-1}(\rho/z). \tag{34}$$

To examine the spectrum in detail, let us consider the hamiltonian defined in Eq. (6). For the axially symmetric chiral field in Eq. (23), this hamiltonian commutes with the third component of the grand spin operator K_3 and the time-reversal operator T. These are specifically,

$$K_3 = L_3 + \frac{1}{2}\sigma_3 + \frac{1}{2}m_{\rm w}\tau_3\,, \tag{35}$$
$$T = i\gamma_1\gamma_3 \cdot i\tau_1\tau_3 C \tag{36}$$

where L_3, σ_3, and τ_3 are respectively the third component of orbital angular momentum, spin, and isospin operator, and C is a charge conjugation operator. The parity operator is defined by $P = \gamma_0$ for odd B, and $P = \gamma_0\tau_3$ for even B. Thus, the eigenvalues of the hamiltonian can be specified by the magnitude of K_3 and the parity $\pi = \pm$. We have $K_3 = 0, \pm1, \pm2, \pm3, \cdots$ for odd B, and $K_3 = \pm\frac{1}{2}, \pm\frac{3}{2}, \pm\frac{5}{2}, \pm\frac{7}{2}, \cdots$ for even B. Since the hamiltonian is invariant under time reverse, the states of $+K_3$ and $-K_3$ are degenerate in energy.

From Fig. 1 it can be seen that for $m_{\rm w} = 2$, the bound states diving into negative region are doubly degenerate with $K_3 = \pm\frac{1}{2}$. Thus putting $N_c = 3$ valence quarks on each of the bound levels, we have the $B = 2$ soliton solution. Similary, for $m_{\rm w} = 3$, we obtained the spectrum of $K_3 = \pm1^-$(double degeneracy) and $K_3 = 0^+$ states diving into negative-energy region which corresponds to the $B = 3$ soliton solution. For $m_{\rm w} = 4$, the spectrum of $K_3 = \pm\frac{1}{2}^+$ and $K_3 = \pm\frac{3}{2}^-$(both doubly degenerate) states dive into negative region, having the $B = 4$ soliton solution. For $m_{\rm w} = 5$, the spectrum of $K_3 = \pm2^+$(double), $K_3 = \pm1^-$(double) and $K_3 = 0^+$ states dive into negative-energy region, having the $B = 5$ soliton solution. These results confirm that the baryon number of the soliton is identified with the number of diving levels occupied by N_c valence quarks. It is interesting that the degeneracy which occurs due to symmetry of the chiral field reduces the number of states, making large shell gaps. This observation indicates that degeneracy in the valence quark spectra contributes to make classical energies of the soliton solutions lower. In fact,

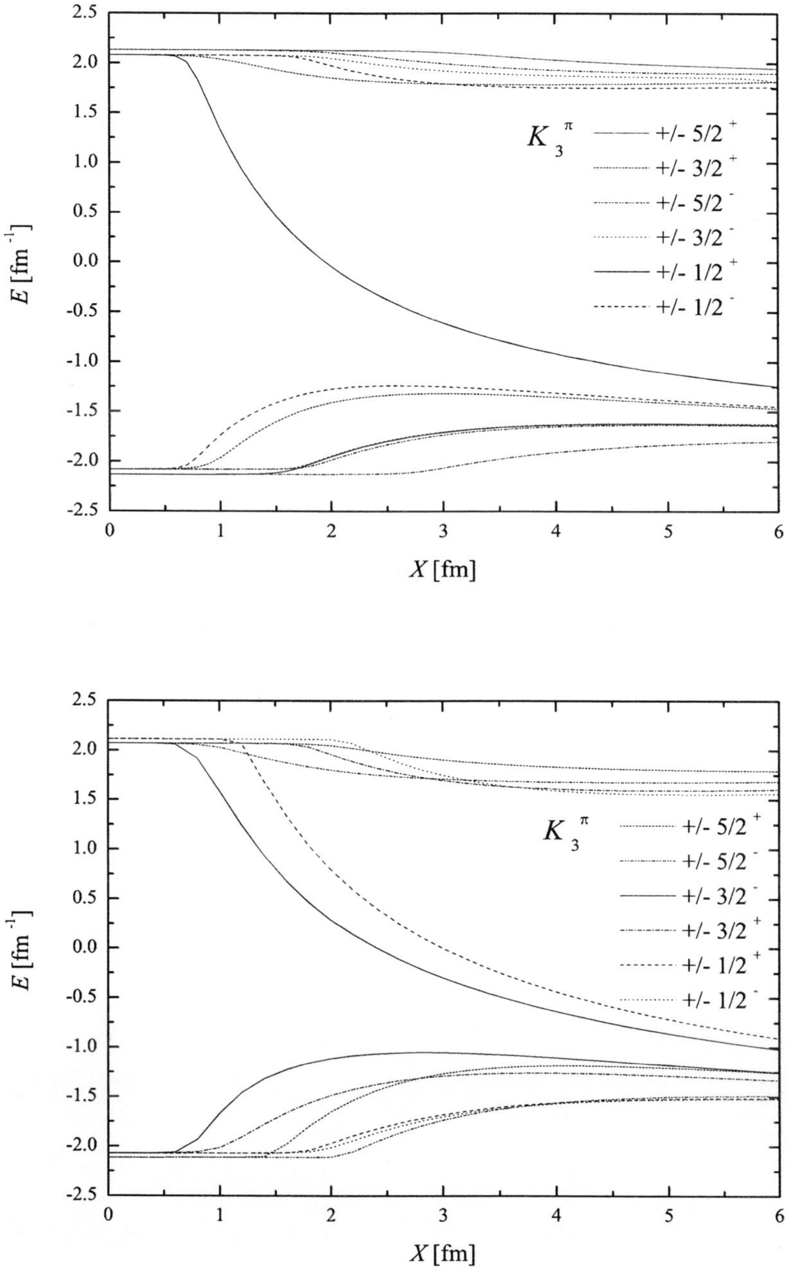

Figure 1: Spectra of the quark orbits of $B = 2$ and $B = 4$ with axially symmetric ansatz (23), as a function of the soliton size parameter X [21].

Table 1: The classical observable, Mass (in MeV), the mean radius of toroid and the root mean square radius (in fm) for $B = 1 \sim 5$.

B			E_{val}			E_{vac}	E_{static}	$\langle\rho\rangle$	$\sqrt{\langle r^2\rangle}$
1	173					674	1192		0.785
2	173	173				1166	2204	0.672	0.821
3	173	173	298			1561	3493	0.659	0.854
4	106	106	232	232		2727	4753	0.971	1.140
5	145	145	319	319	409	2537	6543	1.048	1.225

our $B = 2$ solution which is considered to be the minimum energy soliton from the study of the $B = 2$ skyrmion provides the maximum degeneracy in spectra. It will be interesting to study minimum solutions from this point of view.

The baryon number density is defined by the zeroth component of the baryon current [9],

$$b(\boldsymbol{x}) = \frac{1}{N_c}\langle\bar{\psi}\gamma_0\psi\rangle = b_{\text{val}}(\boldsymbol{x}) + b_{\text{vac}}(\boldsymbol{x}) \tag{37}$$

where

$$\begin{aligned}
b_{\text{val}}(\boldsymbol{x}) &= \frac{1}{N_c}\sum_i \int d\varphi\, \phi_i^\dagger(\rho,\varphi,z)\phi_i(\rho,\varphi,z) \\
b_{\text{vac}}(\boldsymbol{x}) &= \frac{1}{N_c}\Big[\sum_\mu \mathcal{N}(E_\mu)\,\text{sgn}(E_\mu)\int d\varphi\, \phi_\mu^\dagger(\rho,\varphi,z)\phi_\mu(\rho,\varphi,z) \\
&\quad - \sum_\mu \mathcal{N}(E_\mu^{(0)})\,\text{sgn}(E_\mu^{(0)})\int d\varphi \phi_\mu^{(0)\dagger}(\rho,\varphi,z)\phi_\mu^{(0)}(\rho,\varphi,z)\Big].
\end{aligned}$$

A contour plot of the baryon number density for each baryon number is shown in Fig. 4. It can be seen that they have toroidal in shape.

The mean radius $\langle\rho\rangle$ is given by

$$\langle\rho\rangle = \langle\rho\rangle_{\text{val}} + \langle\rho\rangle_{\text{vac}} \tag{38}$$

where

$$\begin{aligned}
\langle\rho\rangle_{\text{val}} &= \frac{1}{m_{\text{w}}}\sum_i \int \rho d\rho dz d\varphi \rho\phi_i^\dagger(\rho,\varphi,z)\phi_i(\rho,\varphi,z)\,, \\
\langle\rho\rangle_{\text{vac}} &= \frac{1}{m_{\text{w}}}\sum_\mu \mathcal{N}(E_\mu)\,\text{sgn}(E_\mu)\int \rho d\rho dz d\varphi \rho\phi_\mu^\dagger(\rho,\varphi,z)\phi_\mu(\rho,\varphi,z)\,.
\end{aligned}$$

The root mean square radius is given by

$$\sqrt{\langle r^2\rangle} = \sqrt{\langle r^2\rangle_{\text{val}} + \langle r^2\rangle_{\text{vac}}} \tag{39}$$

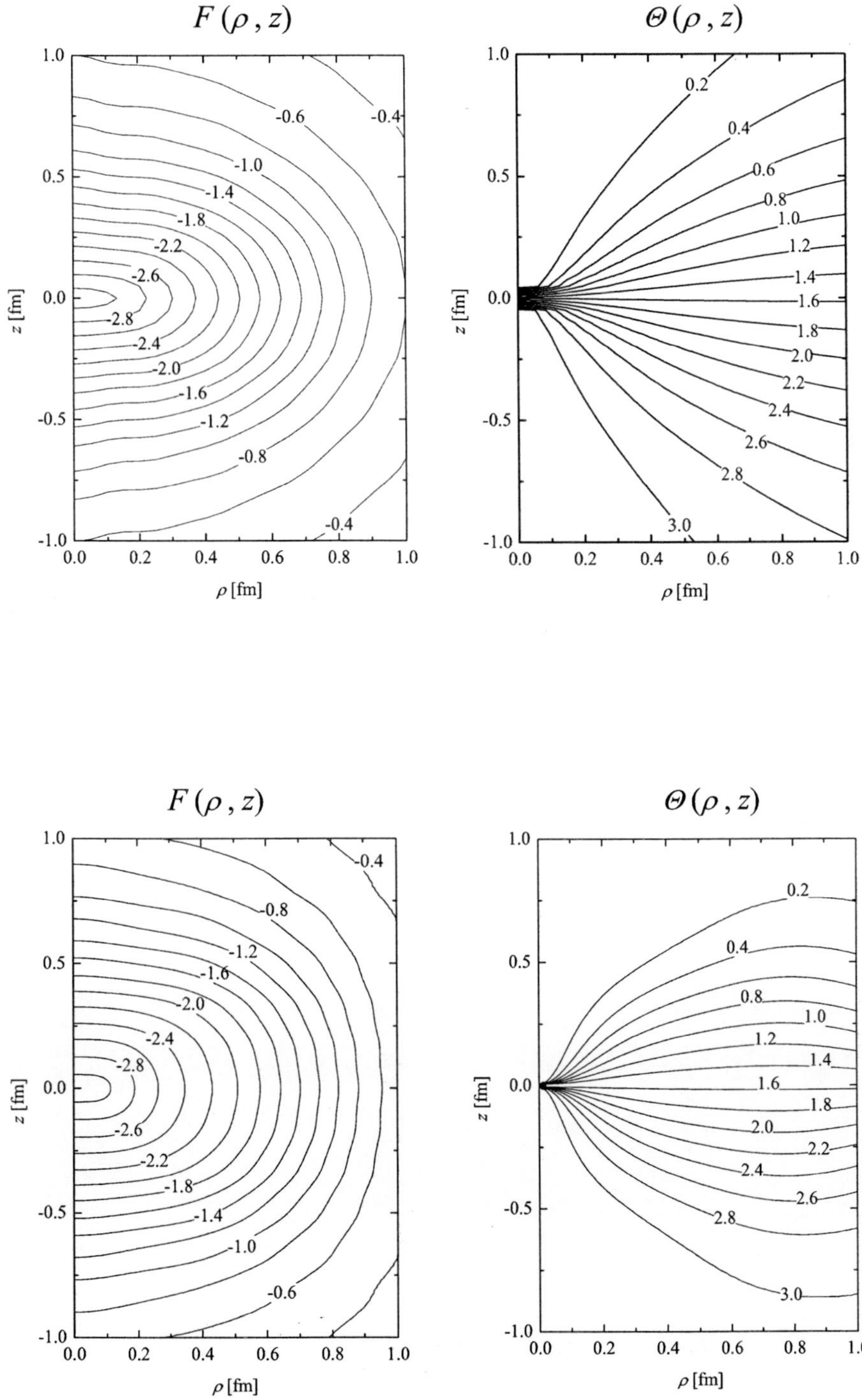

Figure 2: Contour plot of the profile functions $F(\rho, z), \Theta(\rho, z)$, of $B = 2, 3$ with axial symmetry [21].

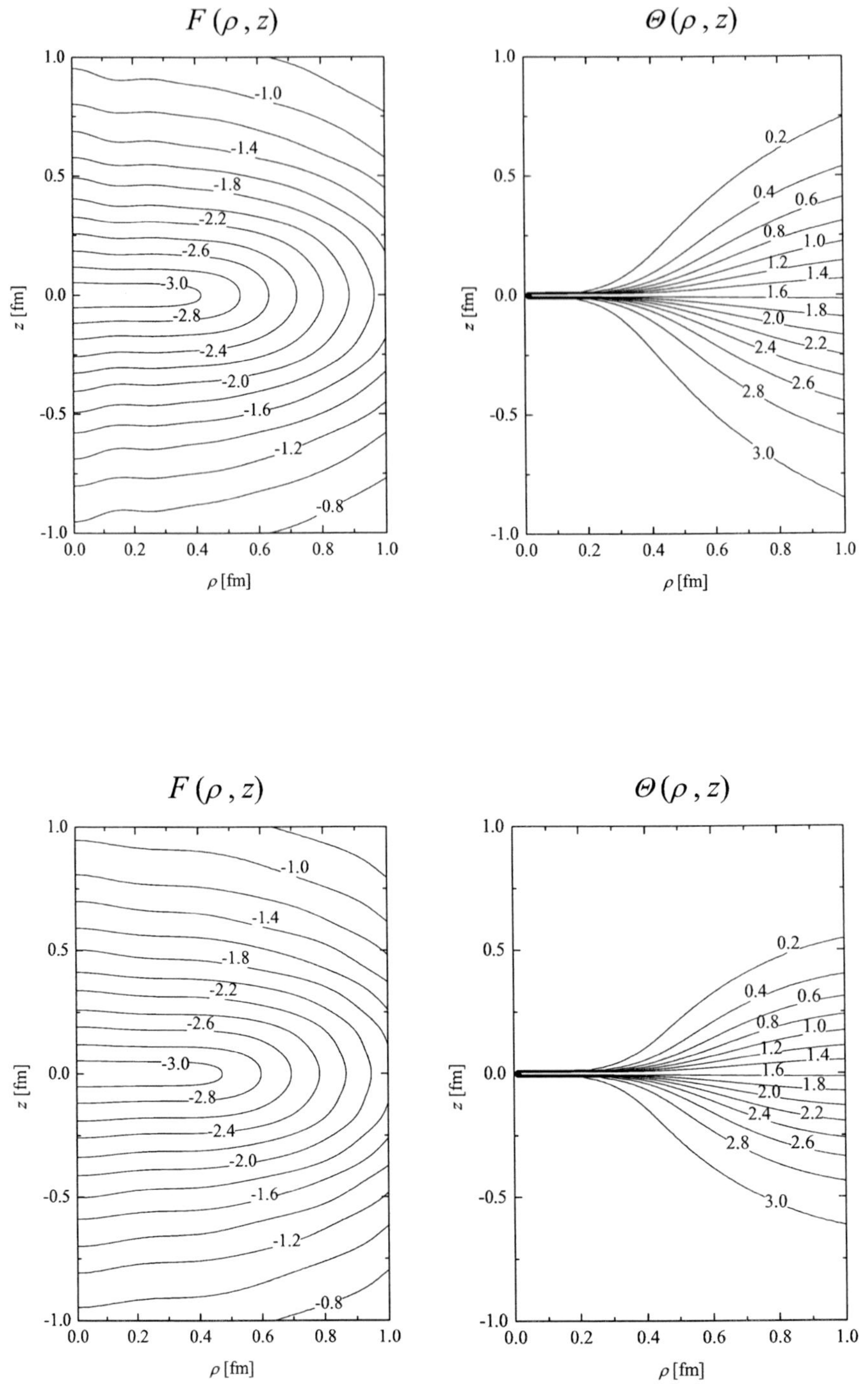

Figure 3: Contour plot of the profile functions $F(\rho, z), \Theta(\rho, z)$, of $B = 4, 5$ with axial symmetry [21].

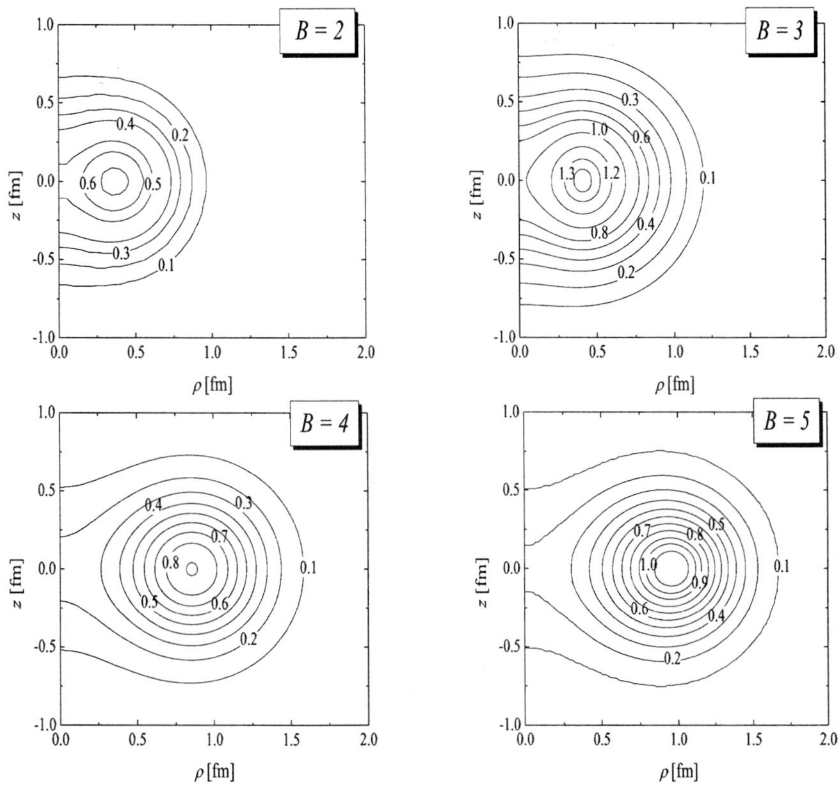

Figure 4: Contour plot of the baryon number densities $b(\boldsymbol{x})$ [fm^{-3}] (37) with axial symmetry [21].

where

$$\langle r^2\rangle_{\rm val} = \frac{1}{m_{\rm W}}\sum_i \int \rho d\rho dz d\varphi(\rho^2+z^2)\phi_i^\dagger(\rho,\varphi,z)\phi_i(\rho,\varphi,z)\,,$$

$$\langle r^2\rangle_{\rm vac} = \frac{1}{m_{\rm W}}\sum_\mu \mathcal{N}(E_\mu)\,{\rm sgn}(E_\mu)\int \rho d\rho dz d\varphi(\rho^2+z^2)\phi_\mu^\dagger(\rho,\varphi,z)\phi_\mu(\rho,\varphi,z)\,.$$

These values for each baryon number are shown in Table 1.

3.2 Multi-Winding Number Configurations with Polyhedral Symmetries

In the Skyrme model it is known that minimal energy configurations with $B \geq 3$ have discrete crystal-like symmetries [16] rather than axisymmetry. We expect that configurations of the CQSM inherits the same discrete symmetry as skyrmions. However, it is too complicated to perform a numerical computation if one imposes such discrete symmetries directly on the chiral fields. Thus Houghton, Manton and Sutcliffe proposed remarkable ansatz for the chiral fields, rational map ansatz [18]. According to this ansatz, the chiral fields are expressed in a rational map as

$$U(r, z) = \exp(iF(r)\hat{\boldsymbol{n}}_R \cdot \boldsymbol{\tau}) , \tag{40}$$

where

$$\hat{\boldsymbol{n}}_R = \frac{1}{1+|R(z)|^2}(2\mathrm{Re}[R(z)], 2\mathrm{Im}[R(z)], 1-|R(z)|^2)$$

and $R(z)$ is the rational map. The complex coordinate z is given by $z = \tan(\theta/2)e^{i\varphi}$ via stereographic projection.

Rational maps are maps from $CP(1)$ to $CP(1)$ (equivalently, from S^2 to S^2) classified by winding number. In Ref. [18] Houghton *et al.* showed that $B = N$ skyrmions can be well-approximated by rational maps with winding number N. The rational map with winding number N possesses $(2N+1)$ complex parameters whose values can be determined so as to realize minimal energies within assumed symmetries of the skyrmion. We shall use this ansatz for the background chiral fields in the CQSM. Their explicit forms of the map are [18, 38]

$$\begin{aligned}
R_3 &= \frac{\sqrt{3}iz^2-1}{z(z^2-\sqrt{3}i)}, \\
R_4 &= \frac{z^4+2\sqrt{3}iz^2+1}{z^4-2\sqrt{3}iz^2+1}, \\
R_5 &= \frac{z(z^4+3.94z^2+3.07)}{3.07z^4-3.94z^2+1}, \\
R_6 &= \frac{z^4+0.16i}{z^2(0.16z^4i+1)}, \\
R_7 &= \frac{7/\sqrt{5}z^6-7z^4-7/\sqrt{5}z^2-1}{z(z^6+7/\sqrt{5}z^4+7z^2-7/\sqrt{5})}, \\
R_8 &= \frac{z^6-0.14}{z^2(0.14z^6+1)}, \\
R_9 &= \frac{z(-3.38-11.19iz^4+z^8)}{1-11.19iz^4-3.38z^8}, \\
R_5^* &= \frac{z(z^4-5)}{-5z^4+1}, \\
R_9^* &= \frac{5i\sqrt{3}z^6-9z^4+3i\sqrt{3}z^2+1-1.98z^2(z^6-i\sqrt{3}z^4-z^2+i\sqrt{3})}{z^3(-z^6-3i\sqrt{3}z^4+9z^2-5i\sqrt{3}-1.98z(-i\sqrt{3}z^6+z^4+i\sqrt{3}z^2-1))}, \\
R_{17} &= \frac{17z^{15}-187z^{10}+119z^5-1}{z^2(z^{15}+119z^{10}+187z^5+17)} .
\end{aligned} \tag{41}$$

Field equations for the chiral fields can be obtained by demanding that the total energy in Eq. (14) be stationary with respect to variation of the profile function $F(r)$,

$$\frac{\delta}{\delta F(r)}E_{\mathrm{static}} = 0 ,$$

Table 2: Mass spectra for $B = 1 - 9, 17$ also for some excited states $B = 5^*, 9^*$(in MeV). The data for $B = 2$ are taken from Ref. [14]. The ratio of the mass E_{static} to $B \times E_{\text{static}}^{(B=1)}$ are compared to that of the Skyrme model [18].

B	$E_{\text{val}}^{(i)}$									E_{field}	E_{static}	$E_{\text{static}}/BE_{\text{static}}^{(B=1)}$	
												Ours	Skyrme
1	173									674	1192	1.00	1.00
2	173	173								1166	2204	0.92	0.95
3	210	210	210							1633	3522	0.98	0.96
4	144	146	146	146						2628	4378	0.92	0.92
5	123	131	131	139	210					3265	5467	0.92	0.93
6	120	124	150	150	206	206				3740	6603	0.92	0.92
7	115	120	120	120	166	166	166			4554	7478	0.90	0.90
8	97	97	115	120	139	139	203	203		5229	8565	0.90	0.91
9	69	101	104	104	107	166	166	179	179	6046	9573	0.89	0.906
17	83	95	95	95	153	156	157	173	175				
	177	178	179	192	194	194	196	196		10586	18650	0.93	0.88
5*	157	157	157	232	232					2874	5680	0.95	1.00
9*	99	105	105	121	142	142	210	210	210	5700	9742	0.91	0.91

which produces

$$S(r)\sin F(r) = P(r)\cos F(r), \tag{42}$$

where

$$S(r) = N_c \sum_{\mu} \big(n_\mu \theta(E_\mu) + \text{sign}(E_\mu)\mathcal{N}(E_\mu)\big)\langle \mu|\gamma^0 \delta(|x| - r)|\mu\rangle\,, \tag{43}$$

$$P(r) = N_c \sum_{\mu} \big(n_\mu \theta(E_\mu) + \text{sign}(E_\mu)\mathcal{N}(E_\mu)\big)\langle \mu|i\gamma^0\gamma^5 \hat{\boldsymbol{n}}_R \cdot \boldsymbol{\tau}\delta(|x| - r)|\mu\rangle\,. \tag{44}$$

The baryon density $b(\boldsymbol{x})$ is defined by

$$b(\boldsymbol{x}) = \frac{1}{N_c}\langle \bar{\psi}\gamma_0\psi\rangle = b_{\text{val}}(\boldsymbol{x}) + b_{\text{field}}(\boldsymbol{x}), \tag{45}$$

where

$$\begin{aligned} b_{\text{val}}(\boldsymbol{x}) &= \sum_i b_{\text{val}}^{(i)}(\boldsymbol{x}) = \frac{1}{N_c}\sum_i \phi_i(\boldsymbol{x})^\dagger \phi_i(\boldsymbol{x})\,, \\ b_{\text{field}}(\boldsymbol{x}) &= \frac{1}{N_c}\Big[\sum_\mu \text{sign}(E_\mu)\mathcal{N}(E_\mu)\phi_\mu(\boldsymbol{x})^\dagger\phi_\mu(\boldsymbol{x}) \\ &- \sum_\mu \text{sign}(E_\mu^{(0)})\mathcal{N}(E_\mu^{(0)})\phi_\mu^{(0)}(\boldsymbol{x})^\dagger\phi_\mu^{(0)}(\boldsymbol{x})\Big]\,. \end{aligned} \tag{46}$$

To examine the shell structure of the quarks, we evaluate the radial density for the ith valence quark $\rho^{(i)}(r)$ in which the angular degrees of freedom are integrated out, via,

$$\rho^{(i)}(r) = \int d\varphi \int \sin\theta d\theta\; b_{\text{val}}^{(i)}(r,\theta,\varphi) \tag{47}$$

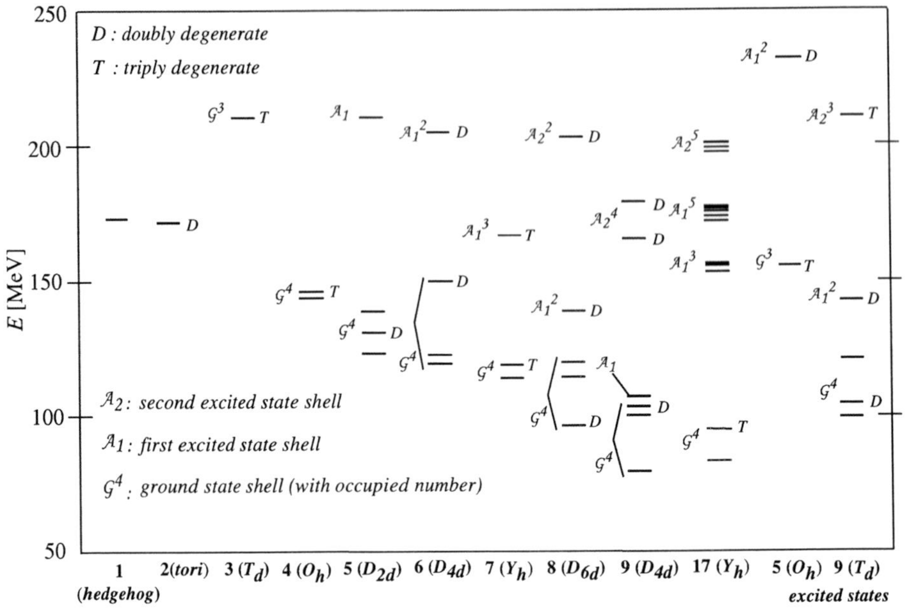

Figure 5: Valence quark spectra for $B = 1 - 9, 17$ [20].

with the baryon number

$$B = \sum_i \int drr^2 \rho^{(i)}(r)\,. \tag{48}$$

The profile functions for $B = 3-9, 17$ are plotted in Fig. 7. In Table 2 are the results for the valence quark levels as well as the vacuum sea contributions. The valence quark spectra show various degenerate patterns depending on the background configuration. The results of the ratio of the mass $E_{\rm static}$ to $B \times E_{\rm static}^{B=1}$ and comparison to those of the Skyrme model are also in Table. 2. They are qualitatively in agreement. Using Eq. (46), we estimated the baryon number density (see Fig. 8). As expected the density inherits the same symmetry as the corresponding skyrmion.

The valence quark spectra for various B are shown in Fig. 5. These results strongly suggest the existence of shell structure for the valence quarks. The spectra show (i) four fold degeneracy of the ground state labeled by $\mathcal{G}$ and various degenerate pattern for excited levels labeled by $\mathcal{A}_1, \mathcal{A}_2, \cdots$, (ii) a large energy gap between the ground state $\mathcal{G}$ and the first excited level $\mathcal{A}_1$. Small dispersions of the spectra are observed in the results. In some cases they are caused by the finite size effect of the basis (ex. $B = 4$). Growing the size $r_{\rm max}$ and increasing the number of the basis, more accurate degeneracy will be attained.

In Fig. 6 are the results of $\rho^{(i)}(r)$ for $B = 3 - 9, 17$. The behaviour of the density near the origin confirms the existence of three shells ($\mathcal{G}, \mathcal{A}_1, \mathcal{A}_2$). $\mathcal{G}$ behaves like "S-wave" and

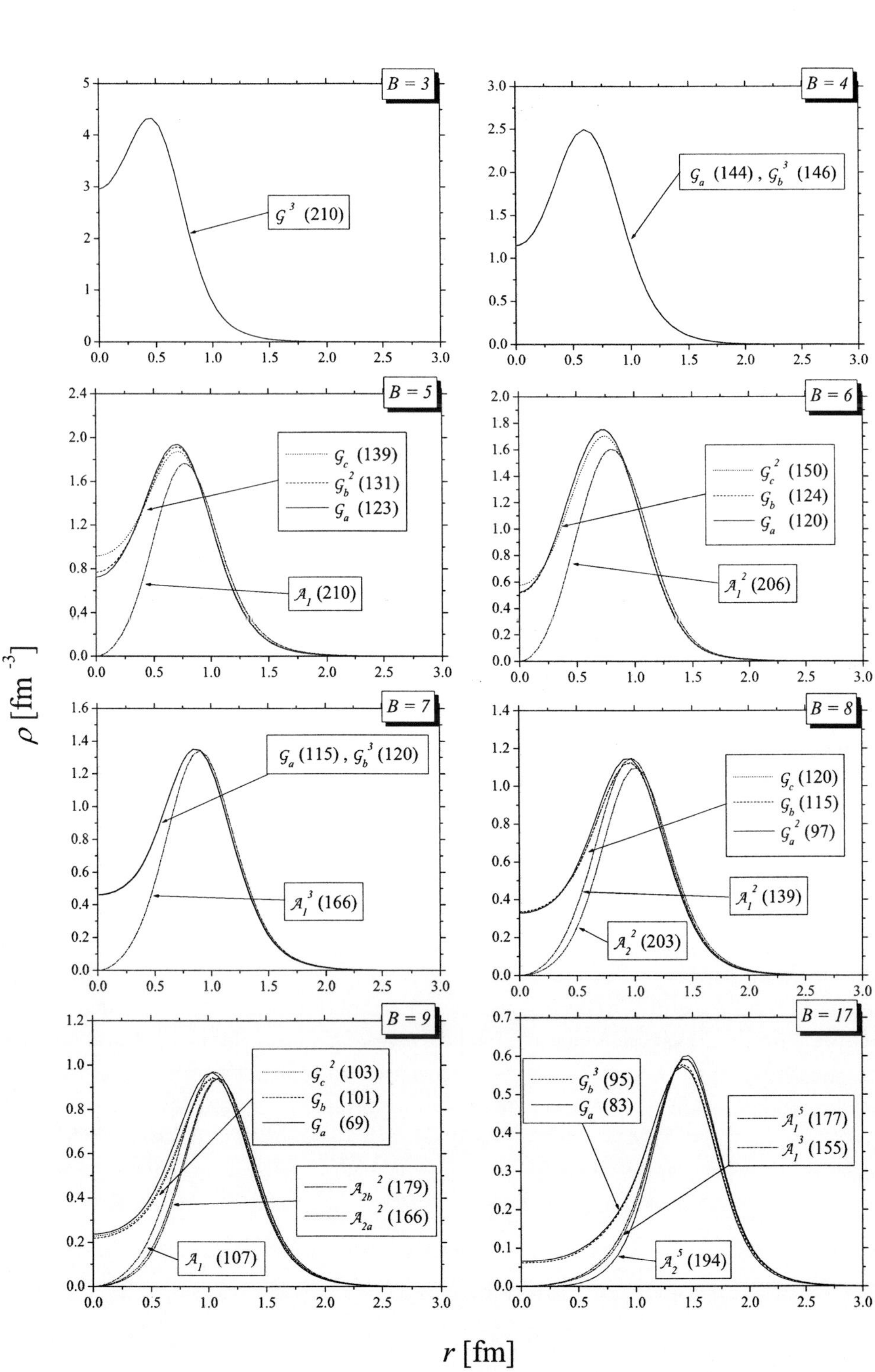

Figure 6: Angular averaged baryon densities of ith valence quarks $\rho^{(i)}(r)$ of $B = 3-9, 17$, with the occupation number and the eigenvalue (in MeV) [20].

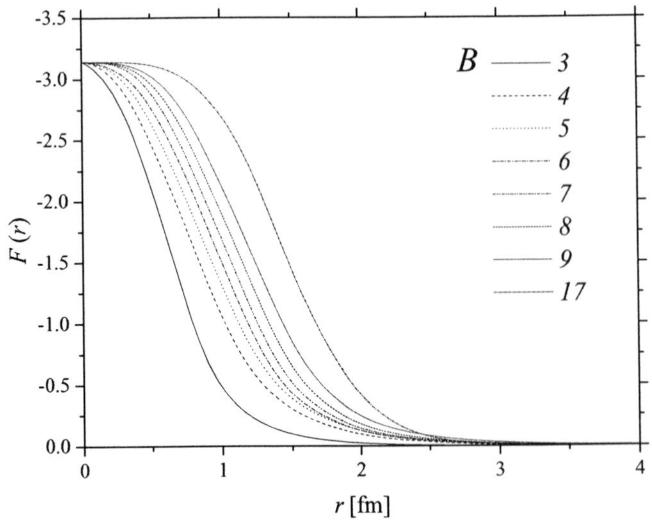

Figure 7: Self-consistent profile functions for $B = 3 - 9, 17$ in the rational map ansatz calculations [20].

others like "P-,D-wave" in a hydrogen-like atom. However most of the densities are nearly on the same surface and very small (not zero) near the origin. The plateau in the density observed at the center of the nucleus [39] can not be attained in our solutions. Therefore one may need to employ the multi-shell ansatz [40] even in the case of light nuclei.

The solutions that we obtained here are totally classical one. To reach the physical observations, one need to take into account the quantum corrections. Especially, the absolute mass tends to be much higher than the real nucleon mass due to a lack of the Casimir effects– the loop corrections of the order of $O(N_c^0)$. For $B = 1$, some attempts for this subject have been done in the Skyrme model [41, 42, 43] and in the hyblid quark soliton model [44]. In both cases the calculated values properly reduce the large classical energy to the physical nucleon mass. For $B > 1$, because of the lower symmetry of the chiral fields, the estimations of these effects are tedious task. Only one work for the $B = 2$ chiral soliton is reported in Ref. [45] in which the configuration is restricted to $SO(3)$. Since the quantum corrections strongly depend on B it is difficult to extract any information of the binding energy from our classical solutions.

3.3 Classical Dynamics of Slowly Moving $B = 2$ Solitons

So far our concern has been restricted to static solutions, we shall now extend our scope to cover dynamics. Once a multi-soliton solution is obtained, one can study the low-energy dynamics of the multi-solitons following the geodesic approximation [46]. The existence of multi-soliton solutions means that there is no net force between the solitons. Manton showed that the multi-BPS monopole solution is a consequence of the cancellation of the

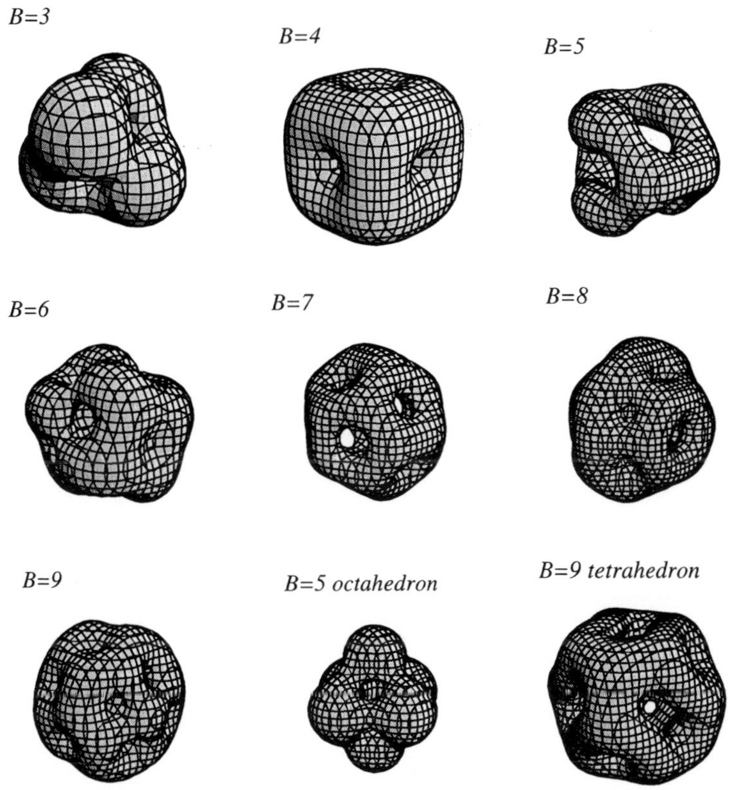

Figure 8: Surface plot of the baryon number densities $b(\boldsymbol{x})$ (45) for $B = 3 - 9$ and the excited states $B = 5^*, 9^*$ [20].

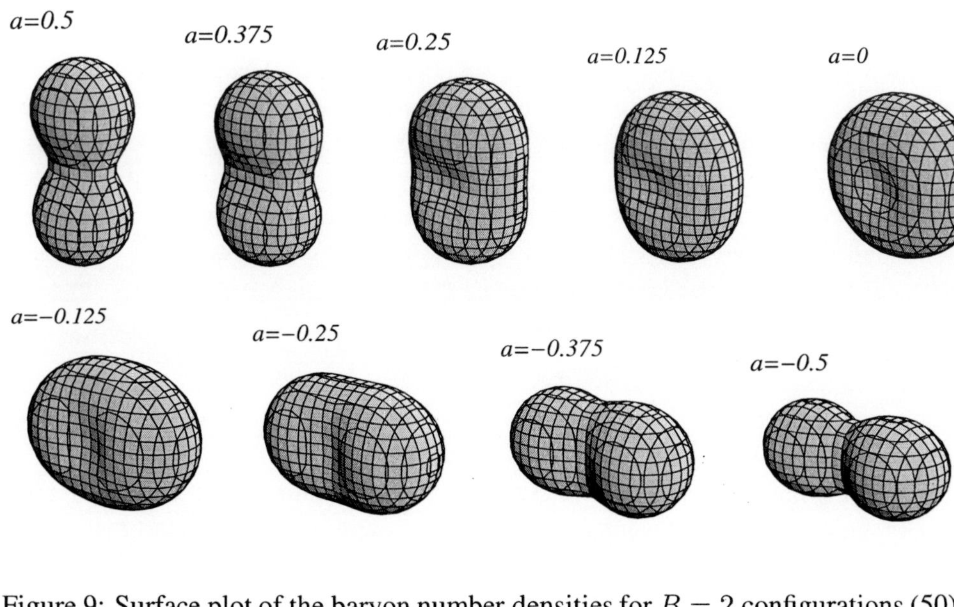

Figure 9: Surface plot of the baryon number densities for $B = 2$ configurations (50) [49].

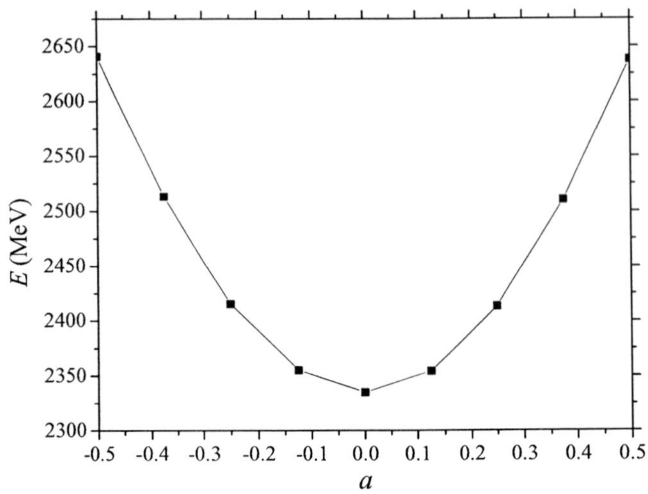

Figure 10: Total energies of the $B = 2$ configurations [49].

repulsive force by the exchange of gauge bosons and the attractive force by exchange of Higgs bosons [47]. The cancellation between the forces can be attained when the solution saturates the Bogomol'nyi bound [48]. For the Skyrme model, one finds

$$E \geq \frac{12\pi^2 f_\pi}{e} B \tag{49}$$

where e is a dimensionless constant which can be fixed experimentally. It is known that skyrmions do not saturate this bound and have slightly higher energies.

If the solitons which are far apart initially are given some impact parameter and start to move slowly towards each other, they will trace a path close to the valley of the potential energy without climbing the wall of the potential barrier. The valley is just the parameter space of the soliton solution which is called a moduli space. If the solitons move slowly, or adiabatically, it will be a good approximation to consider their dynamics only on the moduli space (truncation). Then the time evolution of the solitons is approximately the geodesics on the moduli space.

In the $B = 2$ rational map ansatz, by imposing axisymmetry we get

$$R_2^* = \frac{z^2 - a}{-az^2 + 1} \tag{50}$$

where a is a parameter of a geodesic in the moduli space with $-1 \leq a \leq 1$. For $a = 0$, one recovers the minimal energy toroidal configuration. Letting the parameter a be time-dependent, we can examine the adiabatic time evolution of initially far-apart two solitons, that is,

$$U(r, z; a) = U(r, z; a(t)). \tag{51}$$

The numerical solutions for various values of a are shown in Fig. 9 [49]. It is interesting that the famous 90° scattering is observed as in the BPS monopole [50] and $CP(1)$ solitons [51]. However, as can be seen in the energy level shown in Fig. 10, the solutions have higher energies for larger a, which means the moduli motion is not adiabatic since energy costs to separate the two solitons. This is due to the fact that the chiral solitons does not saturate the so called Bogomol'nyi bound. Thus if the two solitons are initially far apart, they will coalesce to form a bound state, forming a torus shape but no scattering.

4 Symmetry and the Degeneracy of the Quarks

The bunch of valence spectra due to the potential with discrete symmetries has been observed in the study of heavier nuclear systems. In Ref. [52], the valence spectra are highly degenerate because the deformation of the spherically symmetric shell produces large shell gaps. Thus the nuclei can be considered to be more stable than the spherical one. As discussed in Ref. [18], the group theory should predict the level structure of pion fluctuations. However, our problem is more complicated due to the presence of quarks. Before discussing it in detail, let us show how the shell deformation is related to the degeneracy of the spectrum.

In general, if an eigenequation given by

$$H\psi_\mu = E_\mu \psi_\mu \tag{52}$$

is invariant under a symmetric operation $\hat{R} \in \hat{g}$, the equation transforms as

$$\hat{R} H \psi_\mu = H(\hat{R}\psi_\mu) = E_\mu \psi_\mu \,. \tag{53}$$

Therefore the states $\{\psi_\mu, \hat{R}\psi_\mu\}$ are degenerate in energy with E_μ. The set of d_μ eigenfunctions $\{\psi_i^{(\mu)}\}(i = 1, \ldots, d_\mu)$ belonging to a given eigenvalue E_μ will provide the basis for an irreducible representation of the group $\hat{g}$ of the hamiltonian [53]:

$$\hat{R}\psi_j^{(\mu)} = \sum_i \psi_i^{(\mu)} D_{ij}^{(\mu)}(\hat{R}) \,. \tag{54}$$

The Dirac equation

$$(i\gamma^\mu \partial_\mu - M)\psi(x) = 0 \tag{55}$$

is a wave equation for fermions with Lorentz-covariance, i.e. its form has to be invariant under a transition from one inertial system to another one. Let the coordinates of the event be x^μ for an observer A and x'^μ for an observer B. Both coordinates are connected by the Lorentz transformation

$$x' = ax \quad \text{or} \quad (x')^\nu = a^\nu_\mu x^\mu \,. \tag{56}$$

The equation are invariant under this transformation as

$$(i\gamma^\mu \partial_\mu - M)\psi(x) = 0 \;\Leftrightarrow\; (i\gamma'^\mu \partial_\mu - M)\psi'(x') = 0 \,. \tag{57}$$

The Dirac fields are hence transformed via

$$\psi'(x') = \psi'(\hat{a}x) \equiv \hat{S}(\hat{a})\psi(x) = \hat{S}(\hat{a})\psi(\hat{a}^{-1}x') \tag{58}$$

and also

$$\psi(x) = \hat{S}^{-1}(\hat{a})\psi'(x') = \hat{S}^{-1}(\hat{a})\psi'(\hat{a}x)\,,$$
$$\psi'(x') = \hat{S}(\hat{a})\psi'(\hat{a}^{-1}x') \;\Rightarrow\; \psi(x) = \hat{S}(\hat{a}^{-1})\psi'(\hat{a}x)\,.$$

One can thus obtain the relation for the transformation operator $\hat{S}(a)$ as

$$\hat{S}(\hat{a}^{-1}) = \hat{S}^{-1}(\hat{a})\,, \tag{59}$$

which reads to

$$(\hat{S}(\hat{a})i\gamma^{\mu}\partial_{\mu}\hat{S}^{-1}(\hat{a}) - M)\psi'(x') = 0\,. \tag{60}$$

In Eq. (56), performing transformation to the coordinates of the system B,

$$\frac{\partial}{\partial x^{\mu}} = \frac{\partial x'^{\nu}}{\partial x^{\mu}}\frac{\partial}{\partial x'^{\nu}} = \hat{a}^{\nu}_{\ \mu}\frac{\partial}{\partial x'^{\nu}} \tag{61}$$

one gets

$$(\hat{S}(\hat{a})i\gamma^{\mu}\hat{S}^{-1}(\hat{a})\hat{a}^{\nu}_{\mu}\partial'_{\nu} - M)\psi'(x') = 0\,. \tag{62}$$

From the covariance of the Dirac equation in Eq. (57), $\hat{S}(\hat{a})$ must have the following property

$$\hat{a}^{\nu}_{\mu}\gamma^{\mu} = \hat{S}^{-1}(\hat{a})\gamma^{\nu}\hat{S}(\hat{a})\,. \tag{63}$$

Let us consider the transformation law for the equation including the Skyrme chiral fields

$$(i\gamma^{\mu}\partial_{\mu} - MU^{\gamma_5}(x))\psi(x) = 0\,. \tag{64}$$

If the chiral fields have a point group symmetry $\hat{G}$ such as

$$U(x') = \hat{G}(\hat{a})U(x)\hat{G}(\hat{a})^{\dagger}\,,\quad (\hat{G}(\hat{a}) \in SU(2)_I)\,, \tag{65}$$

the Dirac equation is invariant under the Lorentz transformation

$$(x')^{\nu} = a^{\nu}_{\mu}x^{\mu} \text{ or } x' = \hat{a}x \tag{66}$$

with

$$x' = \begin{pmatrix} t \\ \boldsymbol{x}' \end{pmatrix},\quad \hat{a} = \begin{pmatrix} 1 & 0 \\ 0 & \hat{\boldsymbol{a}} \end{pmatrix},\quad x = \begin{pmatrix} t \\ \boldsymbol{x} \end{pmatrix}, \tag{67}$$

accompanying a corresponding iso-rotation

$$(i\gamma^{\mu}\partial_{\mu} - MU^{\gamma_5}(x))\psi(x) = 0 \Leftrightarrow (i\gamma^{\nu}\partial'_{\nu} - MU^{\gamma_5}(x'))\psi'(x') = 0\,, \tag{68}$$

with

$$\psi'(x') = \hat{K}(\hat{a})\psi(x) \equiv (\hat{S}(\hat{a}) \times \hat{G}(\hat{a}))\psi(x)\,. \tag{69}$$

The operator $\hat{R}$ corresponding to this rotation is thus defined by

$$\psi'(x) \equiv_{def} \hat{R}\psi(x) = \hat{K}(\hat{a})\psi(\hat{a}^{-1}x)\,. \tag{70}$$

It can be shown that the operator $\hat{R}$ commutes with the hamiltonian using following commutations :

$$\begin{aligned}
U^{\gamma_5}(x)\psi(x) &= \hat{G}^{-1}U^{\gamma_5}(x')\hat{G}\cdot\hat{R}^{-1}\psi'(x') \\
&= \hat{G}^{-1}\hat{S}^{-1}U^{\gamma_5}(x')\psi'(x')\,, \\
\hat{R}\psi(x') = \hat{K}\psi(x) \quad &\to \quad \psi(x) = \hat{K}^{-1}\hat{R}\psi(x')\,, \\
\Rightarrow \hat{K}\hat{K}^{-1}\hat{R}U^{\gamma_5}(x')\psi(x') &= U^{\gamma_5}(x')\hat{R}\psi(x')\,, \\
\therefore\ \ [\hat{R}, U^{\gamma_5}] &= 0\,,
\end{aligned} \tag{71}$$

$$\begin{aligned}
i\gamma^0\gamma^k\partial_k\psi(x) &= i\gamma^0\gamma^k a_k^l\partial_l'\psi(x) \\
&= i\hat{S}^{-1}\gamma^0\gamma^l\hat{S}\partial_l'\psi(x) \\
&= i\hat{S}^{-1}\gamma^0\gamma^l\hat{S}\partial_l'\hat{K}^{-1}\psi'(x') \\
&= i\hat{G}^{-1}\hat{S}^{-1}\gamma^0\gamma^l\hat{S}\partial_l'\psi'(x') \\
\Rightarrow \hat{K}i\gamma^0\gamma^k\partial_k\psi(x) &= i\gamma^0\gamma^k\partial_k'\psi'(x') \\
\therefore\ \ [\hat{R}, i\gamma^0\gamma^k\partial_k] &= 0\,.
\end{aligned} \tag{72}$$

which reads $[\hat{R}, H] = 0$.

Generally speaking, if eigenequation

$$H\phi_k = \epsilon_k\phi_k \tag{73}$$

is invariant in the group G, in terms of the symmetric operation $\hat{R}$ ($\in G$) the equation becomes

$$\hat{R}H\phi_k = H(\hat{R}\phi_k) = \epsilon_k\phi_k\,. \tag{74}$$

Then, ϕ_k and $\hat{R}\phi_k$ are degenerate in energy. Thus, constructing $\hat{R}$ for each symmetry of the hamiltonian, one should be able to deduce the degeneracy structure of the spectra occurring in the valence level.

As an example, let us examine a rotational operation for the hedgehog ansatz given in Eq. (16). The chiral fields exhibit spherical symmetry and its infinitesimal spatial rotation is of the form

$$x'^{\nu} = a^{\nu}_{\mu}x^{\mu} \quad \to \quad x'^{\nu} = x^{\nu} + \epsilon^{\nu}_{\mu}x^{\mu}\,. \tag{75}$$

Introducing the small angle ϵ as

$$\epsilon_3^2 = -\epsilon_2^3 = \epsilon_1\,,\ \ \epsilon_1^3 = -\epsilon_3^1 = \epsilon_2\,,\ \ \epsilon_2^1 = -\epsilon_1^2 = \epsilon_3\,, \tag{76}$$

the Dirac field transforms as

$$\begin{aligned}\psi(a^{-1}x) &= \psi(x^\nu - \epsilon^\nu_\mu x^\mu) \\ &= \psi(x^\nu) + \frac{\partial\psi}{\partial x^\nu}(-\epsilon^\nu_\mu) \\ &= \psi(x^\nu) + \boldsymbol{\epsilon}\cdot(\boldsymbol{x}\times\nabla)\psi(x^\nu) \\ &= (1 + i(\boldsymbol{x}\times\boldsymbol{p})\cdot\boldsymbol{\epsilon})\psi(x^\nu) = \exp[i\boldsymbol{l}\cdot\boldsymbol{\epsilon}]\psi(x^\nu)\,, \end{aligned} \tag{77}$$

where $\boldsymbol{l} = \boldsymbol{x}\times\boldsymbol{p}$ is orbital angular momentum operator. An infinitesimal transformation of the Lorentz and isorotation operators are given by

$$\begin{aligned}&\hat{S} = 1 - \frac{i}{4}\sigma^{\mu\nu}\epsilon_{\mu\nu} = 1 + \frac{i}{2}\boldsymbol{\Sigma}\cdot\boldsymbol{\epsilon}\,,\ \ \boldsymbol{\Sigma} \equiv \begin{pmatrix}\boldsymbol{\sigma} & 0\\ 0 & \boldsymbol{\sigma}\end{pmatrix}, \\ &\hat{I} = 1 + \frac{i}{2}\boldsymbol{\tau}\cdot\boldsymbol{\epsilon}\,. \end{aligned} \tag{78}$$

Thus the transformation of the Dirac field under the Lorentz and isorotation is given by (see Eq. (70))

$$\begin{aligned}\psi'(x) = \hat{R}\psi(x) &= \hat{S}\times\hat{I}\psi(a^{-1}x) \\ &= (1 + \frac{i}{2}\boldsymbol{\Sigma}\cdot\boldsymbol{\epsilon})(1 + \frac{i}{2}\boldsymbol{\tau}\cdot\boldsymbol{\epsilon})(1 + i(\boldsymbol{x}\times\boldsymbol{p})\cdot\boldsymbol{\epsilon})\psi(x) \\ &\sim (1 + i(\frac{1}{2}\boldsymbol{\Sigma} + \frac{1}{2}\boldsymbol{\tau} + \boldsymbol{x}\times\boldsymbol{p})\cdot\boldsymbol{\epsilon})\psi(x) = (1 + i\boldsymbol{K}\cdot\boldsymbol{\epsilon})\psi(x)\,. \end{aligned} \tag{79}$$

This $\boldsymbol{K}$ is a grandspin operator and a good quantum number for the hedgehog hamiltonian.

The point group symmetry has a discrete operation so that it is not possible to perform these infinitesimal transformation analysis. Nevertheless, one can see that the rotation in Eq. (70) produces the degenerate eigenstates. Let us show briefly the derivation of the rotation operator $\hat{R}$ for $B = 3$ tetrahedron and the transformation law for the numerical basis constructed in Eq. (173). The $B = 3$ tetrahedral soliton is characterized by two symmetry operations [18]: $Z_2 \times Z_2$ and T_d. Specifically, $Z_2 \times Z_2$ are characterized by a following Möbius transformation:

$$z \to -\frac{1}{z} \ \Leftrightarrow\ R(z) \to -\frac{1}{R(z)} \equiv R'(z) \tag{80}$$

resultantly,

$$\begin{aligned}&\hat{\boldsymbol{n}}_R(z') = (n_1, n_2, n_3) \\ &\to \hat{\boldsymbol{n}}'_R(z) = \frac{1}{1 + |R(z)|^2}\left(2\mathrm{Re}[R'(z)], 2\mathrm{Im}[R'(z)], 1 - |R'(z)|^2\right) \\ &\qquad = \frac{1}{1 + 1/|R|^2}\left(2\mathrm{Re}[-1/R], 2\mathrm{Im}[-1/R], 1 - |1/R|^2\right) \\ &\qquad = (-n_1, n_2, -n_3)\,. \end{aligned} \tag{81}$$

The transformation operator $\hat{g} \equiv \exp[-i\frac{\pi}{2}\tau_2]$ ensures

$$\begin{aligned} \hat{g}(\boldsymbol{\tau}\cdot\boldsymbol{n}_R)\hat{g}^\dagger &= (-i\tau_2)(\tau_1 n_1 + \tau_2 n_2 + \tau_3 n_3)(i\tau_2) \\ &= -n_1\tau_1 + n_2\tau_2 - n_3\tau_3 \equiv \boldsymbol{\tau}\cdot\boldsymbol{n}'_R\,. \end{aligned} \tag{82}$$

and transforms the chiral field as

$$U(\boldsymbol{x}') = \hat{g}U(\boldsymbol{x})\hat{g}^\dagger\,. \tag{83}$$

T_d transforms the z and $R(z)$ as

$$z \to \frac{iz+1}{-iz+1} \equiv z' \;\Leftrightarrow\; R(z) \to \frac{iR+1}{-iR+1} \equiv R'(z)\,, \tag{84}$$

and hence

$$\begin{aligned} & n_1 + in_2 \\ & \to n'_1 + in'_2 = \frac{2R'}{1+|R'|^2} = \frac{1+i(R+\bar{R})-|R|^2}{1+|R|^2} = n_3 + in_1\,, \\ & n_1 - in_2 \\ & \to n'_1 - in'_2 = \frac{2\bar{R}'}{1+|R'|^2} = \frac{1-i(R+\bar{R})-|R|^2}{1+|R|^2} = n_3 - in_1\,, \\ & n_3 \to n'_3 = \frac{1-|R'|^2}{1+|R'|^2} = \frac{-i(R-\bar{R})}{1+|R|^2} = n_2\,, \end{aligned} \tag{85}$$

which yields

$$\begin{aligned} n'_1\tau_1 + n'_2\tau_2 + n'_3\tau_3 &= \frac{1}{2}(n'_1 + in'_2)(\tau_1 - i\tau_2) + \frac{1}{2}(n'_1 - in'_2)(\tau_1 + i\tau_2) + n'_3\tau_3 \\ &= \frac{1}{2}(n_3 + in_1)(\tau_1 - i\tau_2) + \frac{1}{2}(n_3 - in_1)(\tau_1 + i\tau_2) + n_2\tau_3 \\ &= n_3\tau_1 + n_1\tau_2 + n_2\tau_3\,. \end{aligned} \tag{86}$$

The transformation operator $\hat{h} \equiv \exp[-i\frac{\pi}{3}\frac{1}{\sqrt{3}}(\tau_1+\tau_2+\tau_3)] = (1 - i(\tau_1+\tau_2+\tau_3))/2$ ensures

$$\begin{aligned} \hat{h}(\boldsymbol{\tau}\cdot\boldsymbol{n}_R)\hat{h}^\dagger &= \frac{1}{2}(1-i(\tau_1+\tau_2+\tau_3))(\tau_1 n_1 + \tau_2 n_2 + \tau_3 n_3)\frac{1}{2}(1+i(\tau_1+\tau_2+\tau_3)) \\ &= n_3\tau_1 + n_1\tau_2 + n_2\tau_3 \equiv \boldsymbol{\tau}\cdot\boldsymbol{n}'_R\,, \end{aligned} \tag{87}$$

and transforms the chiral field as

$$U(\boldsymbol{x}') = \hat{h}U(\boldsymbol{x})\hat{h}^\dagger\,. \tag{88}$$

The Lorentz transformation operator $\hat{S}$ and the operators for the chiral fields $\{\hat{g}, \hat{h}\}$

corresponding to the symmetric operations $(x')^\nu = a^\nu_\mu x^\mu$ are given by

$Z_2 \times Z_2$:

$$(a_g)^\nu_\mu = \begin{pmatrix} 1 & 0 & 0 & 0 \\ 0 & -1 & 0 & 0 \\ 0 & 0 & 1 & 0 \\ 0 & 0 & 0 & -1 \end{pmatrix} \Rightarrow \hat{S}_g = i\gamma^0\gamma^5\gamma^2 , \tag{89}$$

T_d :

$$(a_h)^\nu_\mu = \begin{pmatrix} 1 & 0 & 0 & 0 \\ 0 & 0 & 1 & 0 \\ 0 & 0 & 0 & 1 \\ 0 & 1 & 0 & 0 \end{pmatrix} \Rightarrow \hat{S}_h = \exp[i\frac{\pi}{3}\frac{1}{\sqrt{3}}(\sigma_{23} + \sigma_{31} + \sigma_{12})] . \tag{90}$$

The $\hat{R}$ is defined by the direct product of these rotation operators together with the inverse spatial rotation for the spinor such as

$$\hat{R}_g\psi(x) \equiv (\hat{S}_g \times \hat{g})\psi(\hat{a}_g^{-1}x) , \tag{91}$$

$$\hat{R}_h\psi(x) \equiv (\hat{S}_h \times \hat{h})\psi(\hat{a}_h^{-1}x) . \tag{92}$$

Applying these operators to the Kahana-Ripka basis $\phi \equiv \{u, v\}$ (for detail, see Appendix B) we finally obtain the following transformation laws:

$$\hat{R}_g\phi_{KM} = (-1)^{K-M}\phi_{K-M} . \tag{93}$$

$$\hat{R}_h\phi_{00} = \phi_{00} . \tag{94}$$

$$\hat{R}_h \begin{pmatrix} \phi_a \\ \phi_b \\ \phi_c \end{pmatrix} = \begin{pmatrix} 0 & -1 & 0 \\ 0 & 0 & -1 \\ 1 & 0 & 0 \end{pmatrix} \begin{pmatrix} \phi_a \\ \phi_b \\ \phi_c \end{pmatrix} , \tag{95}$$

$$\phi_a \equiv \frac{1}{\sqrt{2}}(\phi_{11} + \phi_{1-1}) , \; \phi_b \equiv i\phi_{10} , \; \phi_c \equiv \frac{i}{\sqrt{2}}(\phi_{11} - \phi_{1-1}) .$$

$$\hat{R}_h \begin{pmatrix} \phi_\xi \\ \phi_\eta \\ \phi_\zeta \\ \phi_u \\ \phi_v \end{pmatrix} = \left(\begin{array}{ccc|cc} 0 & -1 & 0 & 0 & 0 \\ 0 & 0 & 1 & 0 & 0 \\ -1 & 0 & 0 & 0 & 0 \\ \hline 0 & 0 & 0 & -\frac{1}{2} & \frac{\sqrt{3}}{2} \\ 0 & 0 & 0 & -\frac{\sqrt{3}}{2} & -\frac{1}{2} \end{array}\right) \begin{pmatrix} \phi_\xi \\ \phi_\eta \\ \phi_\zeta \\ \phi_u \\ \phi_v \end{pmatrix} , \tag{96}$$

$$\phi_\xi = \frac{i}{\sqrt{2}}(\phi_{21} + \phi_{2-1}) , \; \phi_\eta = \frac{1}{\sqrt{2}}(\phi_{21} - \phi_{2-1}) ,$$

$$\phi_\zeta = \frac{i}{\sqrt{2}}(\phi_{22} - \phi_{2-2}) , \; \phi_u = \phi_{20} , \; \phi_v = \frac{1}{\sqrt{2}}(\phi_{22} + \phi_{2-2}) ,$$

confirming that the $B = 3$ tetrahedron can exhibit triply degenerate spectra.

The numerical computation indicates that the winding number strongly couple the elements with different K and correlated valence spectra occur as a result. As can be seen from the operator K with $B = 2$, the degeneracy of the valence spectra are explained in terms of the shape deformation (symmetry) as well as the winding number of the chiral fields [13]. The four-fold degeneracy of the lowest states may be ascribed to the chiral symmetry $SU(2)_L \times SU(2)_R$ of the hamiltonian. The degenerate structure for $B \geq 3$ will be well understood if symmetric operators of the hamiltonian which consist of the angular momentum, spin, isospin and winding number, are explicitly constructed.

5 Zero Mode Quantization

The solitons that we obtained in the previous section are classical objects and therefore must be quantized to assign definite spin and isospin to them. Quantization of the solitons can be performed semiclassically for their rotational zero modes. For the hedgehog soliton, because of its topological structure, a rotation in isospin space is followed by a simultaneous spatial rotation. For the axially symmetric soliton, there are five rotational zero modes by rotations of iso-degrees of freedom and spatial rotations.

5.1 $SU(2)$ Sector

Let us introduce the dynamically rotated chiral fields around the classical fields U_S [37]:

$$U(\boldsymbol{x},t) = A(t)U_S(\boldsymbol{x}')A(t)^\dagger, \;\; x^{i'} = \Xi^i_j[B(t)]x^j \tag{97}$$

where

$$\Xi^i_j[B(t)] = \frac{1}{2}\mathrm{Tr}[\sigma^i B(t)\sigma_j B(t)^\dagger]\,, \tag{98}$$

and $A(t)$ and $B(t)$ are time-dependent $SU(2)$ matrices generating an iso-rotation and a spatial rotation respectively. By transforming the rotating frame of reference, the Dirac operator with Eq. (97) can be written as

$$\begin{aligned} i\tilde{D} &= i\partial\!\!\!/ - MU^{\gamma_5}(\boldsymbol{x},t) \\ &= A(t)S(t)^\dagger\gamma^0[i\partial_t + i\tilde{\gamma^0}\tilde{\gamma^k}\partial_k{}' - MU_S^{\gamma_5}(\boldsymbol{x}') + iA^\dagger\dot{A} + iS^\dagger\dot{S}]S(t)A(t)^\dagger \end{aligned} \tag{99}$$

where

$$\tilde{\gamma^\mu} = \Lambda^\mu_\nu S\gamma^\nu S^\dagger = \begin{pmatrix} \gamma^0 \\ \gamma^k + (\boldsymbol{r}'\times\dot{\boldsymbol{\theta}})^k\gamma^0 \end{pmatrix}, \tag{100}$$

and $S(t)$ is the rotation operator for the Dirac field and $\boldsymbol{\theta}$ is an angle of the spatial rotation. Note that the gamma matrices $\tilde{\gamma^\mu}$ explicitly depend on the coordinates and do not transform as a contravariant vector [54]. Substituting Eq. (100) into Eq. (99), one obtains

$$i\tilde{D} = A(t)S(t)^\dagger\gamma^0[i\partial_t - H(U_S^{\gamma_5}) + \Omega_A + \Omega_B]S(t)A(t)^\dagger \tag{101}$$

where

$$\Omega_A = iA^\dagger\dot{A} = \frac{1}{2}\Omega^a_A\tau_a\,, \;\; \Omega_B = iS^\dagger\dot{S} + (\boldsymbol{r}\times\boldsymbol{p})\cdot\dot{\boldsymbol{\theta}} = \Omega^a_B J_b \tag{102}$$

with $J_a = 1/2\epsilon_{abc}\gamma^b\gamma^c - i(\boldsymbol{r}\times\nabla)_a$. Ω_A and Ω_B are the angular velocity operators for an isorotation and for a spatial rotation respectively. Under these operations the effective action can be written by

$$\begin{aligned} S_{\text{eff}}(U) &= S_{\text{eff}}(U_S) \\ &- iN_c\,\text{Sp}\left[\log(i\partial_t - H(U_S^{\gamma_5}) + \Omega_A + \Omega_B)\right] - \text{Sp}\left[\log(i\partial_t - H(U_S^{\gamma_5})\right] \\ &\to S_{\text{eff}}(U_S) - \frac{1}{2}iN_c\,\text{Sp}\log dd^\dagger + \frac{1}{2}iN_c\,\text{Sp}\log d_0 d_0{}^\dagger\,. \end{aligned} \tag{103}$$

With the proper-time regularization, it reads

$$S_{\text{eff}}^{\text{reg}}(U) = S_{\text{eff}}^{\text{reg}}(U_S) - \frac{N_c}{2}\int\frac{d\omega}{2\pi}\int_{1/\Lambda^2}^{\infty}\frac{d\tau}{\tau}\,\text{Sp}[e^{-dd^\dagger\tau} - e^{-d_0 d_0{}^\dagger\tau}]\,, \tag{104}$$

where

$$d = i\omega - H(U_S^{\gamma_5}) + \Omega_A + \Omega_B,\, d^\dagger = -i\omega - H(U_S^{\gamma_5}) - \Omega_A - \Omega_B\,, \tag{105}$$

$$dd^\dagger = \omega^2 + H^2 - 2i\omega(\Omega_A + \Omega_B) - [H, \Omega_A + \Omega_B] - (\Omega_A + \Omega_B)^2\,, \tag{106}$$

$$d_0 d_0{}^\dagger = \omega^2 + H^2\,. \tag{107}$$

Assuming that the rotation of the soliton is adiabatic, we shall expand the effective action S_{eff} around the classical solution U_S with respect to the angular momentum velocity Ω_A and Ω_B up to second order [55]

$$\begin{aligned} S_{\text{eff}}^{\text{reg}}(U) &= S_{\text{eff}}^{\text{reg}}(U_S) \\ &+ \frac{1}{2}\sum_{ab}\int dt\big[I_{0,ab}^{AA}\Omega_A^a(t)\Omega_A^b(t) + I_{0,ab}^{AB}\Omega_A^a(t)\Omega_B^b(t) \\ &+ I_{0,ab}^{BA}\Omega_B^a(t)\Omega_A^b(t) + I_{0,ab}^{BB}\Omega_B^a(t)\Omega_B^b(t)\big] \end{aligned} \tag{108}$$

where $I_{0,ab}$'s are the vacuum sea contributions to the moments of inertia defined by

$$I_{0,ab}^{AA} = \frac{1}{8}N_c\sum_{n,m} f(E_m, E_n, \Lambda)\langle n|\tau_a|m\rangle\langle m|\tau_b|n\rangle\,,$$

$$I_{0,ab}^{AB} = \frac{1}{4}N_c\sum_{n,m} f(E_m, E_n, \Lambda)\langle n|\tau_a|m\rangle\langle m|J_b|n\rangle\,,$$

$$I_{0,ab}^{BA} = \frac{1}{4}N_c\sum_{n,m} f(E_m, E_n, \Lambda)\langle n|J_a|m\rangle\langle m|\tau_b|n\rangle\,,$$

$$I_{0,ab}^{BB} = \frac{1}{2}N_c\sum_{n,m} f(E_m, E_n, \Lambda)\langle n|J_a|m\rangle\langle m|J_b|n\rangle$$

with the cutoff function $f(E_m, E_n, \Lambda)$

$$\begin{aligned} f(E_m, E_n, \Lambda) &= -\frac{2\Lambda}{\sqrt{\pi}}\frac{e^{-E_m^2/\Lambda^2} - e^{-E_n^2/\Lambda^2}}{E_m^2 - E_n^2} \\ &+ \frac{\text{sgn}(E_m)\text{erfc}(|E_m|/\Lambda) - \text{sgn}(E_n)\text{erfc}(|E_n|/\Lambda)}{E_m - E_n}\,. \end{aligned} \tag{109}$$

Similarly, for the valence quark contribution to the moments of inertia, we have

$$
\begin{aligned}
I^{AA}_{\mathrm{val},ab} &= \frac{1}{2}N_c \sum_{m\neq \mathrm{val}} \frac{\langle \mathrm{val}|\tau_a|m\rangle\langle m|\tau_b|\mathrm{val}\rangle}{E_m - E_{\mathrm{val}}}\,, \\
I^{AB}_{\mathrm{val},ab} &= N_c \sum_{m\neq \mathrm{val}} \frac{\langle \mathrm{val}|\tau_a|m\rangle\langle m|J_b|\mathrm{val}\rangle}{E_m - E_{\mathrm{val}}}\,, \\
I^{BA}_{\mathrm{val},ab} &= N_c \sum_{m\neq \mathrm{val}} \frac{\langle \mathrm{val}|J_a|m\rangle\langle m|\tau_b|\mathrm{val}\rangle}{E_m - E_{\mathrm{val}}}\,, \\
I^{BB}_{\mathrm{val},ab} &= 2N_c \sum_{m\neq \mathrm{val}} \frac{\langle \mathrm{val}|J_a|m\rangle\langle m|J_b|\mathrm{val}\rangle}{E_m - E_{\mathrm{val}}}\,.
\end{aligned}
\tag{110}
$$

The total moments of inertia are then given by the sum of the vacuum and valence as $I^{AA}_{ab} = I^{AA}_{\mathrm{val},ab} + I^{AA}_{0,ab}$.

Finally, the effective lagrangian is obtained as

$$
L = -E_{\mathrm{static}} + \frac{1}{2} I^{AA}_{ab}\Omega^a_A\Omega^b_A + I^{AB}_{ab}\Omega^a_A\Omega^b_B + \frac{1}{2} I^{BB}_{ab}\Omega^a_B\Omega^b_B\,. \tag{111}
$$

Theoretically, these moments of inertia can be computed using the eigenstates of Eq. (8). However, due to the difference of the boundary conditions between the initial and final states of the matrix element, the moments of inertia acquire nonzero values with vanishing pion fields. To overcome this problem, we make the following replacement [56]:

$$
\begin{aligned}
\langle n|J_a|m\rangle &\rightarrow \langle n|[H(U_S^{\gamma_5}), J_a]|m\rangle/(E_n - E_m) \\
&= \langle n|[MU_S^{\gamma_5}, l_a]|m\rangle/(E_n - E_m)
\end{aligned}
\tag{112}
$$

where $l_a = -i(\boldsymbol{r}\times\nabla)_a$. Unless the hamiltonian explicitly depend on the coordinates, the numerator vanishes with vanishing pion fields. The spurious contributions to the moment of inertia can be removed in this way.

From axial symmetry of the system, following relations are derived

$$
\begin{aligned}
&I_{ij} = 0,\ \ i\neq j;\ I^{AB}_{11} = I^{AB}_{22} = I^{BA}_{11} = I^{BA}_{22} = 0\,, \\
&I^{AA}_{11} = I^{AA}_{22},\ I^{BB}_{11} = I^{BB}_{22}\,,\ I^{BB}_{33} = m^2 I^{AA}_{33},\ I^{AB}_{33} = I^{BA}_{33} = -m_{\mathrm{w}} I^{AA}_{33}.
\end{aligned}
\tag{113}
$$

The quantization conditions for the collective coordinates, $A(t)$ and $B(t)$, define a body-fixed isospin operator $\boldsymbol{K}$ and a body-fixed angular momentum operator $\boldsymbol{L}$ as

$$
I^{AA}_{ab}\Omega^b_A + I^{AB}_{ab}\Omega^b_B \rightarrow -\,\mathrm{tr}\left(A\frac{\tau_a}{2}\frac{\partial}{\partial A}\right) \equiv -K_a\,, \tag{114}
$$

$$
I^{BA}_{ab}\Omega^b_A + I^{BB}_{ab}\Omega^b_B \rightarrow \mathrm{tr}\left(\frac{\sigma_a}{2}B\frac{\partial}{\partial B}\right) \equiv -L_a. \tag{115}
$$

These are related to the usual coordinate-fixed isospin operator I_a and coordinate-fixed angular momentum J_a operator by transformations,

$$
I_a = -\Xi^b_a[A(t)]K_b,\quad J_a = -\Xi^b_a[B(t)]^T L_b. \tag{116}
$$

Table 3: Moments of inertia (in MeV^{-1}).

B		Valence	Sea	Total	B		Valence	Sea	Total
2	I_{11}^{AA}	0.00773	0.00363	0.01136	4	I_{11}^{AA}	0.01408	0.00959	0.02366
	I_{11}^{BB}	0.01141	0.00464	0.01605		I_{11}^{BB}	0.04272	0.01245	0.05517
	I_{33}^{AA}	0.00429	0.00125	0.00554		I_{33}^{AA}	0.01172	0.00074	0.01246
3	I_{11}^{AA}	0.01231	0.00280	0.01511	5	I_{11}^{AA}	0.02786	0.00716	0.03502
	I_{11}^{BB}	0.02174	0.00384	0.02558		I_{11}^{BB}	0.12124	0.01112	0.13236
	I_{33}^{AA}	0.00594	0.00027	0.00622		I_{33}^{AA}	0.01368	0.00007	0.01375

To estimate the quantum energy corrections, let us introduce the basis functions of the spin and isospin operators which were inspired from the cranking method for nuclei [39, 36],

$$\langle A, B | i i_3 k_3, j j_3 l_3 \rangle = \frac{\sqrt{(2i+1)(2j+1)}}{8\pi^2} D^{i\ *}_{i_3 k_3}(A) D^{j\ *}_{j_3, -m_w k_3}(B)$$

where D is the Wigner rotation matrix. Then, we find the quantized energies of the soliton as

$$\begin{aligned} E &= E_{\text{static}} \\ &+ \frac{1}{2I_{11}^{AA}} i(i+1) + \frac{1}{2I_{11}^{BB}} j(j+1) + \frac{1}{2}\left(\frac{1}{I_{33}^{AA}} - \frac{1}{I_{11}^{AA}} - \frac{m_{\mathrm{w}}^2}{I_{11}^{BB}}\right) k_3^2 \end{aligned} \tag{117}$$

where $i(i+1), j(j+1)$ and k_3 are eigenvalues of the Casimir operators $\boldsymbol{I}^2$ and $\boldsymbol{J}^2$, and the operator $\boldsymbol{K}_3$, respectively.

In Table 3 are the results of our calculation of moments of inertia, I_{11}^{AA}, I_{11}^{BB} and I_{33}^{AA}, with $B = 2-5$. It is instructive to compare our results with the Skyrme model [37] where $U_{11} = 0.0104$, $V_{11} = 0.0163$ and $U_{33} = 0.00709$ which are correspondingly our I_{11}^{AA}, I_{11}^{BB} and I_{33}^{AA}. They are qualitatively in good agreement.

5.2 Finkelstein-Rubinstein Constraints

If a multi-skyrmion describes atomic nuclei upon quantization, it has to be quantized as a boson or as a fermion whether B is even or odd. This requirement is implemented in the form of Finkelstein-Rubinstein (FR) constraints [57]. For highly nonlinear theory enough to possess soliton solutions a consideration of continuity reduces to a concept of distinct topological sector. Finkelstein and Rubinstein state that an equally primitive concept as continuity is between multi-valued quantized systems which can possess state functions double-valued under 2π rotation, and those which cannot. Both concepts are ascribed to homotopy of the map between the physical space and the configuration space. Indeed for the $SU(2)$ chiral soliton solution to exist, the physical space must be compactified to S^3 which defines a topological charge characterized by an integer

$$\pi_3(S^3) = n\,. \tag{118}$$

The FR constraints arise when the space-time is suitably compactified as

$$\pi_4(SU(2)) = Z_2\,, \tag{119}$$

which takes values only -1 or $+1$. This allows us to quantize the solitons as either a fermion or a boson.

The FR constraints for the rational map ansatz was constructed in Ref. [22] and Ref. [58] and applied to predict the ground states of skyrmions up to $B = 22$. In this section, we shall apply the FR constraints for the rational map ansatz directly to our axially symmetric multi-skyrmions and obtain their ground states.

Following the notation in Ref. [58], let g be a rotation by α around $\boldsymbol{n}$ followed by an isorotation by β around $\boldsymbol{N}$. Then the FR constraints can be defined by

$$\exp(-i\alpha\boldsymbol{n}\cdot\boldsymbol{J})\exp(-i\beta\boldsymbol{N}\cdot\boldsymbol{I})\psi = \chi_{FR}(g)\psi \tag{120}$$

where

$$\chi_{FR}(g) = \begin{cases} 1 & \text{if contractible} \\ -1 & \text{otherwise.} \end{cases} \tag{121}$$

and, $\boldsymbol{J}$ and $\boldsymbol{I}$ are space-fixed spin and isospin operators respectively. ψ is the wave function which transforms under a tensor product of rotations and isorotations. In particular, a closed loop is noncontractible for odd B and contractible for even B, which is consistent with spin statistics. Consequently, quantum numbers I and J are half-integers for odd B and integers for even B.

In order to construct the ground states for a given baryon number B, let us define $N(L(\alpha,\beta))$ as a homotopy invariant for a loop L generated by rotations by α and isorotations by β. Then, for the axially symmetric rational map of degree B, it is given by Ref. [58]

$$N(L(\alpha,\beta)) = \frac{B}{2\pi}(B\alpha - \beta)\,. \tag{122}$$

It can be shown that N (mod 2) determines if the loop is contractible or not in the same sense as B (mod 2). Therefore, N (mod 2) gives the FR constraints for each generator of the symmetry group of the rational map.

The axially symmetric rational map with degree B is given by

$$R(z) = \frac{1}{z^B}\,. \tag{123}$$

There are two symmetric generators for this rational map. One is a rotation by α followed by an isorotation by $\beta = B\alpha$. Substituting it into Eq. (122), one obtains $N(L(\alpha, B\alpha)) = 0$. The FR constraints for this loop is thus given by

$$e^{-i\pi(L_3 - BK_3)}\psi = \psi\,. \tag{124}$$

where L_3, K_3 are the third component of the body-fixed angular momentum and the isospin operators which are related with the space-fixed operators by orthogonal transformations (116). The other symmetry is C_2 with transformation

$$z \to \frac{1}{z}\,, \quad R(z) \to \frac{1}{R(z)}\,. \tag{125}$$

This corresponds to $\alpha = \beta = \pi$ and hence $N(L(\pi,\pi)) = B(B-1)/2$. The FR constraints for this loop is

$$e^{-i\pi(L_1+K_1)}\psi = (-1)^{B(B-1)/2}\psi\,. \tag{126}$$

In the following we construct the ground states consistent with the derived FR constraints (124) and (126) for $B = 2-5$ with axial symmetry.

- $B = 2$
  We find the FR constraints

$$e^{-i\pi(L_3-2K_3)}\psi = \psi \tag{127}$$
$$e^{-i\pi(L_1+K_1)}\psi = -\psi\,. \tag{128}$$

  This gives the ground state as $|J, L_3\rangle\,|I, K_3\rangle = |1,0\rangle\,|0,0\rangle$. This is in agreement with the ground state ${}^{1}_{1}\mathrm{H}^{+}$ (deuteron).

- $B = 3$
  We find the FR constraints

$$e^{-i\pi(L_3-3K_3)}\psi = \psi \tag{129}$$
$$e^{-i\pi(L_1+K_1)}\psi = -\psi \tag{130}$$

  This gives the ground state as $\left|\frac{5}{2},\frac{3}{2}\right\rangle\left|\frac{1}{2},\frac{1}{2}\right\rangle$. The ground state (and first "excited" state) of $B = 3$ have $(I, J) = \left(\frac{1}{2},\frac{1}{2}\right)$ (${}^{3}_{2}\mathrm{He}^{+}$, ${}^{3}_{1}\mathrm{H}^{+}$). Then, one can not identify our soliton with these observed isodoublet nuclei.

- $B = 4$
  We find the FR constraints

$$e^{-i\pi(L_3-4K_3)}\psi = \psi \tag{131}$$
$$e^{-i\pi(L_1+K_1)}\psi = \psi\,. \tag{132}$$

  This gives the ground state as $|0,0\rangle\,|0,0\rangle$. The ground state ${}^{4}_{2}\mathrm{He}^{+}$ has $(I, J) = (0,0)$. Then our soliton can be identified as the "α particle".

- $B = 5$
  We find the FR constraints

$$e^{-i\pi(L_3-5K_3)}\psi = \psi \tag{133}$$
$$e^{-i\pi(L_1+K_1)}\psi = \psi\,. \tag{134}$$

  This gives the ground state as $|J\rangle\,|I\rangle = \left|\frac{7}{2},\frac{5}{2}\right\rangle\left|\frac{1}{2},\frac{1}{2}\right\rangle$. This is not in agreement with the observed nuclei of $B = 5$; $(I, J) = \left(\frac{1}{2},\frac{3}{2}\right)$ (${}^{5}_{2}\mathrm{He}^{+}$, ${}^{5}_{3}\mathrm{Li}^{+}$).

Table 4: $B = 2$, mass spectrum up to $i, j \leq 3, k_3 \leq 1$.

Classification	(i, j, k_3)	Parity	Mass[MeV]
$NN(^3S_1)$	$(0, 1, 0)$	$+$	2264
$NN(^1S_0)$	$(1, 0, 0)$	$+$	2290
$N\Delta(^3P_2)$	$(1, 2, 1)$	$-$	2399
$N\Delta(^5S_2)$	$(1, 2, 0)$	$+$	2477
$N\Delta(^3S_1)$	$(2, 1, 0)$	$+$	2528
$\Delta\Delta(^7S_3)$	$(0, 3, 0)$	$+$	2576
$\Delta\Delta(^1S_0)$	$(3, 0, 0)$	$+$	2730
$\Delta\Delta(^5P_3)$	$(2, 3, 1)$	$-$	2762

Thus, we conclude that for even B, the axially symmetric solitons are possible candidates of the ground states of B atomic nuclei as is the case of the deuteron and $^4_2$He while for odd B they emerge only as excited states.

The results of the quantized energy of the axially symmetric soliton solutions with $B = 2, 4$ are shown in Table 4 and 5. The study of the Finkelstein-Rubinstein constraints indicates that the axially symmetric solution with even B has the same quantum number as the physically observed nuclei. Some of the states may be observed in experiments. Specifically, in the $B = 2$, we obtained the $I = 0, J = 1$ ($^3S_1$: deuteron) and $I = 1, J = 0$ ($^1S_0$) solutions. The energy of $^3S_1$ is lower than the $^1S_0$ because $I_{11}^{BB} > I_{11}^{AA}$ (see Table 3). The order is in agreement with the experimental observations. For $B = 4$, the quantum number of the ground state $I = 0, J = 0$ coincide with the observation. In experiment, the lowest excited state also has $I = 0, J = 0$, which unfortunately can not be explained within our scheme. This state can be interpreted as $^3_2$He+n bound state or $^3_1$H+p resonance state rather than the resonance of single $^4_2$He. Our formulation for the multi-soliton is based on the simgle bag-like picture. We expect that for some resonance states one needs advanced formulation including multi-fragments such as $B = (3 + 1), (2 + 2)$. We observed $I = 0, J = 2$ with positive parity as a first excited state and this channel should emerge as a higher resonance (roughly 28 MeV from the ground state) in experiment. For odd B, the constraint of C_2 in Eq. (126) seems to assure the validity of the ansatz. Indeed, it provides the ground state as $I = J = 1/2$ for $B = 3$ and as $I = 1/2, J = 3/2$ for $B = 5$, which coincide with physical observations. This seems to make sense since in the minimal energy configurations with discrete symmetries, the solutions tend to have $I = J = 1/2$ due to their shell-like structure. However, unfortunately the constraint in Eq. (124) forbits such states. Consequently, the axially symmetric solitons with odd B emerge only as excited states. The resultant lowest state is $E = 3657$ MeV with $I = 1/2, J = 5/2$ for $B = 3$, and $E = 6591$ MeV with $I = 1/2, J = 7/2$ for $B = 5$. Experimentally, no possible candidate of the state $I = 1/2, J = 7/2$ are found.

As stated in Sec. 2, in the soliton approach the absolute mass always tends to be higher due to the lack of the Casimir effects. Therefore the total energies is overestimated around $0.5 \sim 1$ GeV in our calculation. The one-loop corrections and vacuum effects should be properly subtracted in order to estimate physical mass of the solutions. For $B = 1$, the

Table 5: $B = 4$, mass spectrum up to $i \leq 3, j \leq 5, k_3 \leq 1$.

Classification	(i, j, k_3)	Parity	Mass[MeV]
${}^4N({}^1S_0)$	$(0, 0, 0)$	$+$	4753
${}^4N({}^5S_2)$	$(0, 2, 0)$	$+$	4807
${}^2N\,{}^2\Delta({}^7P_4)$	$(0, 4, 1)$	$-$	4808
${}^4N({}^3S_1)$	$(1, 1, 0)$	$+$	4813
${}^4N({}^1S_0)$	$(2, 0, 0)$	$+$	4879
${}^3N\,\Delta({}^7S_3)$	$(1, 3, 0)$	$+$	4904
${}^4N({}^5S_2)$	$(2, 2, 0)$	$+$	$4934_{.2}$
${}^2N\,{}^2\Delta({}^7S_4)$	$(0, 4, 0)$	$+$	$4934_{.3}$
${}^2N\,{}^2\Delta({}^9P_4)$	$(2, 4, 1)$	$-$	4935
$N\,{}^3\Delta({}^9P_5)$	$(1, 5, 1)$	$-$	4941
${}^3N\,\Delta({}^3S_1)$	$(3, 1, 0)$	$+$	5025
$N\,{}^3\Delta({}^9S_5)$	$(1, 5, 0)$	$+$	5046
${}^2N\,{}^2\Delta({}^9S_4)$	$(2, 4, 0)$	$+$	5061
${}^3N\,\Delta({}^7S_3)$	$(3, 3, 0)$	$+$	5115
$N\,{}^3\Delta({}^9P_5)$	$(3, 5, 1)$	$-$	5152
$N\,{}^3\Delta({}^9S_5)$	$(3, 5, 0)$	$+$	5278

obtained mass of the nucleon the delta is $E_N = 1260$ MeV and $E_\Delta = 1505$ MeV respectively because of the lack of Casimir effects. With these values, the ground states of the $B = 2 \sim 4$ exhibit bound states but the ground state of the $B = 5$ does not. For $B = 2$, the state $I = 0, J = 3$ (${}^7S_3$) has a large binding energy. The search of this resonance is interesting subject and also a thorough analysis of the Casimir effects for the toroidal solitons are much desired for a stability argument.

5.3 $SU(3)$ Sector

The $SU(2)_L \times SU(2)_R$-invariant lagrangian has a natural extension to $SU(3)$ sector including strange quarks. The $SU(3)_L \times SU(3)_R$-invariant lagrangian in the chiral quark soliton model is given by

$$\mathcal{L} = \bar{\psi}(i\partial\!\!\!/ - MU^{\gamma_5} - \hat{m})\psi \tag{135}$$

where $U^{\gamma_5}(x) = e^{i\gamma_5\pi_a(x)\lambda_a/f_\pi}$ and λ_a are the usual Gell-Mann matrices with $\lambda_0 = \sqrt{\frac{2}{3}}\,\hat{1}$. In order to estimate the effects of the symmetry breaking of $SU(3)$ explicitly, we introduce the current quark mass matrix

$$\hat{m} = \text{diag}(m_0, m_0, m_s) = m_0\hat{1} + \Delta m(1 - \sqrt{3}\lambda_8)/3 \tag{136}$$

where $\Delta m \equiv m_s - m_0$ is the mass difference between the strange- and the u-,d-quark.

Within the proper-time regularization scheme, we fix our parameters by the input parameters, pion mass m_π, kion mass m_K, and the pion decay constant f_π in the following formulae [59, 60]

$$\frac{N_c M^2}{4\pi^2}\int_0^\infty \frac{d\tau}{\tau}\phi(\tau)e^{-\tau\bar{M}^2} = f_\pi^2 \tag{137}$$

$$m_0\frac{N_c M}{2\pi^2 f_\pi^2}\int_0^\infty \frac{d\tau}{\tau^2}\phi(\tau)e^{-\tau\bar{M}^2} = m_\pi^2 \tag{138}$$

$$(m_0+\frac{\Delta m}{2})\frac{N_c M}{2\pi^2 f_\pi^2}\int_0^\infty \frac{d\tau}{\tau^2}\phi(\tau)e^{-\tau\bar{M}^2} = m_K^2 \tag{139}$$

where $\bar{M} = M + m_0$. For the damping function $\phi(\tau)$ we introduce the two cut-off Λ_1, Λ_2 and the parameter c via

$$\phi(\tau) = c\theta(\tau - 1/\Lambda_1^2) + (1-c)\theta(\tau - 1/\Lambda_2^2). \tag{140}$$

Using the values $M = 400$ MeV , $m_0 = 6$ MeV , $f_\pi = 93$ MeV, $m_\pi = 138$ MeV and Eqs. (137),(138) we obtain the parameter set $\{c = 0.76, \Lambda_1 = 433.8\,\text{MeV}, \Lambda_2 = 1512.4\,\text{MeV}\}$. By $m_K = 496$ MeV and Eq. (139), we determine $m_s = 149$ MeV.

For the extension of the three flavor soliton with $B = 2$, we follow the usual collective coordinate approach for the hedgehog ansatz with $B = 1$ [61, 59]. Within the collective coordinate approach, the extension to $SU(3)$ is performed by trivial embedding [62]:

$$U(\boldsymbol{x}, t) = A(t)\begin{pmatrix} U_0(\Lambda_j^i(t)x^j) & 0 \\ 0 & 1 \end{pmatrix} A^\dagger(t)\,, \tag{141}$$

where $A(t)$ is the time-dependent $SU(3)$ collective rotation matrix and $\Lambda_j^i(t) = \frac{1}{2}\text{Tr}(\tau^i B\tau_j B^\dagger)$ is the spatial rotation matrix. Substituting (141) into the quark determinant and transforming the rotated frame of reference, one obtains [15]:

$$\begin{aligned} iD &= i\partial\!\!\!/ - MU^{\gamma_5}(\boldsymbol{x},t) - \hat{m} \\ &\to A(t)S(t)^\dagger\gamma_0(i\partial_t - H(U_S^{\gamma_5}) - H_{SB} - \Omega_A + \Omega_B)S(t)A(t)^\dagger\,, \end{aligned} \tag{142}$$

where

$$\Omega_A = -iA^\dagger\dot{A} = \frac{1}{2}\Omega_A^a\lambda_a\,, \tag{143}$$

$$\Omega_B = \Omega_B^a(\frac{1}{2}\epsilon_{abc}\gamma^b\gamma^c - i(\boldsymbol{r}\times\nabla)_a) = \Omega_B^a J_a\,, \tag{144}$$

and

$$H_{SB} = A^\dagger(t)\beta\Delta m\frac{1}{3}(1-\sqrt{3}\lambda_8)A(t)\,. \tag{145}$$

$S(t)$ is the rotation operator for the Dirac fields. In the rotating system, the quarks feel the induced Coliolis forces Ω_A, Ω_B and H_{SB}. Ω_A, Ω_B are the angular velocity operators for the right flavor rotation and the spatial rotation. The H_{SB} represents the contribution to the hamiltonian due to the $SU(3)$ symmetry breaking.

We assume that the rotational velocities and the mass difference are relatively small and the expansion in powers of Ω_A, Ω_B and Δm is rapidly convergent. Expanding the quark determinant and the (valence quark) Green function in terms of the collective angular velocity up to the second order and the quark mass difference of the first order. The effective action (5) with Eq. (142) can be rewrite as

$$S_{\text{eff}} = -iN_c \text{Sp} \log iD \to S_{\text{eff}}(U_S) + S_R + S_I \,. \tag{146}$$

S_R, S_I mean the real and imaginary contribution to the action, which are explicitly written as

$$S_R = -\frac{i}{2}N_c[\text{Sp} \log dd^\dagger - \text{Sp} \log d_0 d_0{}^\dagger] \,, \quad S_I = -\frac{i}{2}N_c \text{Sp} \log d(d^\dagger)^{-1} \,, \tag{147}$$

where $d = i\partial_t - H(U_S) - H_{\text{SB}} - \Omega_A + \Omega_B$. The expansion for the real part with proper-time regularization form yields

$$\begin{aligned} S_R &\to -\int \frac{d\tau}{\tau} \text{Sp}\phi(\tau)[e^{-dd^\dagger\tau} - e^{-d_0 d_0{}^\dagger\tau}] \\ &\to \int dt \Big[-\frac{1}{2}\gamma_0(1 - D_{88}) + \frac{1}{2}I^{AA}_{ab,0}\Omega^a_A\Omega^b_A - I^{AB}_{ab,0}\Omega^a_A\Omega^b_B + \frac{1}{2}I^{BB}_{ab,0}\Omega^a_B\Omega^b_B \Big] \end{aligned} \tag{148}$$

and

$$\gamma_0 = -\Delta m \frac{N_c}{3} \sum_{i=1,2} \sum_n c_i \, \text{sgn}(E_n) \text{erfc}\Big(\frac{|E_n|}{\Lambda_i}\Big) \langle n|\gamma_0|n\rangle \,, \tag{149}$$

$$I^{AA}_{ab,0} = \frac{N_c}{8} \sum_{i=1,2} \sum_{m,n} c_i \, f(E_m, E_n, \Lambda_i) \langle n|\lambda_a|m\rangle \langle m|\lambda_b|n\rangle \,, \tag{150}$$

$$I^{AB}_{ab,0} = \frac{N_c}{4} \sum_{i=1,2} \sum_{m,n} c_i \, f(E_m, E_n, \Lambda_i) \langle n|\lambda_a|m\rangle \langle m|J_b|n\rangle \,, \tag{151}$$

$$I^{BB}_{ab,0} = \frac{N_c}{2} \sum_{i=1,2} \sum_{m,n} c_i \, f(E_m, E_n, \Lambda_i) \langle n|J_a|m\rangle \langle m|J_b|n\rangle \,, \tag{152}$$

where $c_1 = c, c_2 = 1 - c$ and the cutoff function $f(E_m, E_n, \Lambda)$ is defined in Eq. (109). For the estimation of the imaginary part of the action, we need the following manipulation,

$$\begin{aligned} S_I &= -\frac{i}{2}N_c \text{Sp} \log d(d^\dagger)^{-1} \\ &= \int dt \int \frac{d\omega}{2\pi} \text{Tr} \log \Big(\frac{w - H(U_S) - H_{SB} - \Omega_A + \Omega_B}{w - H(U_S) - H_{SB} + \Omega_A - \Omega_B} \Big) \\ &= \int dt \int \frac{d\omega}{2\pi} \int_{-1}^{1} d\lambda \text{Tr} \Big(\frac{-\Omega_A + \Omega_B}{w - H(U_S) - H_{SB} - \lambda(\Omega_A - \Omega_B)} \Big) \,. \end{aligned}$$

With replacement $\omega - \lambda(\Omega_A - \Omega_B) \to \omega$ and analytical continuation for ω, we obtain

$$\begin{aligned} &= \int dt \int \frac{d\omega}{2\pi} \int_{-1}^{1} d\lambda \text{Tr} \Big(\frac{-\Omega_A + \Omega_B}{w - H(U_S) - H_{SB}} \Big) \\ &= 2i \int \frac{d\omega'}{2\pi} \text{Sp} \frac{(\Omega_A - \Omega_B)(H(U_S) + H_{SB})}{\omega'^2 + (H(U_S) + H_{SB})^2} \,, \quad \omega = i\omega' \,. \end{aligned} \tag{153}$$

Introducing the following proper-time regularization,

$$\frac{1}{\omega'^2 + (H(U_S) + H_{SB})^2} \rightarrow \int_{1/\Lambda^2}^{\infty} d\tau \exp[-\tau(\omega'^2 + (H(U_S) + H_{SB})^2)] \tag{154}$$

and expand up to first order, we finally obtain the form

$$S_I \rightarrow -\int dt \left[\frac{\sqrt{3}}{2} B[U_S]\Omega_A^8 + K_{ab,0}^A D_{8a}\Omega_A^b + K_{ab,0}^B D_{8a}\Omega_B^b \right] \tag{155}$$

and

$$K_{ab,0}^A = \Delta m \frac{N_c}{4\sqrt{3}} \sum_{m,n} F(E_m, E_n, \Lambda) \langle n|\beta\lambda_a|m\rangle\langle m|\lambda_b|n\rangle \,, \tag{156}$$

$$K_{ab,0}^B = \Delta m \frac{N_c}{2\sqrt{3}} \sum_{m,n} F(E_m, E_n, \Lambda) \langle n|\beta\lambda_a|m\rangle\langle m|J_b|n\rangle \,. \tag{157}$$

The moments of inertia K_{ab}^A, K_{ab}^B are derived from imaginary part of the effective action thus need no regularization. The "cut-off" function $F(E_m, E_n, \Lambda)$ becomes

$$F(E_m, E_n, \Lambda \rightarrow \infty) = \frac{\mathrm{sgn}(E_m) - \mathrm{sgn}(E_n)}{E_m - E_n} \,. \tag{158}$$

Also the valence quark contributions for the moments of inertia read

$$\gamma_{\text{val}} = \Delta m \frac{2N_c}{3} \langle \text{val}|\gamma_0|\text{val}\rangle \,, \tag{159}$$

$$K_{ab,\text{val}}^A = \Delta m \frac{N_c}{\sqrt{3}} \sum_{n \neq \text{val}} \frac{\langle \text{val}|\beta\lambda_a|n\rangle\langle n|\lambda_b|\text{val}\rangle}{E_n - E_{\text{val}}} \,, \tag{160}$$

$$K_{ab,\text{val}}^B = \Delta m \frac{2N_c}{\sqrt{3}} \sum_{n \neq \text{val}} \frac{\langle \text{val}|\beta\lambda_a|n\rangle\langle n|J_b|\text{val}\rangle}{E_n - E_{\text{val}}} \,, \tag{161}$$

$$I_{ab,\text{val}}^{AA} = \frac{N_c}{2} \sum_{n \neq \text{val}} \frac{\langle \text{val}|\lambda_a|n\rangle\langle n|\lambda_b|\text{val}\rangle}{E_n - E_{\text{val}}} \,, \tag{162}$$

$$I_{ab,\text{val}}^{AB} = N_c \sum_{n \neq \text{val}} \frac{\langle \text{val}|\lambda_a|n\rangle\langle n|J_b|\text{val}\rangle}{E_n - E_{\text{val}}} \,, \tag{163}$$

$$I_{ab,\text{val}}^{BB} = 2N_c \sum_{n \neq \text{val}} \frac{\langle \text{val}|J_a|n\rangle\langle n|J_b|\text{val}\rangle}{E_n - E_{\text{val}}} \,. \tag{164}$$

Finally we can construct the following form for the effective lagrangian

$$\begin{aligned} L &= -E_{\text{static}} - \frac{\sqrt{3}}{2} B[U_S]\Omega_A^8 - \frac{1}{2}\gamma(1 - D_{88}) \\ &- K_{ab}^A D_{8a}\Omega_A^b - K_{ab}^B D_{8a}\Omega_B^b \\ &+ \frac{1}{2} I_{ab}^{AA}\Omega_A^a\Omega_A^b - I_{ab}^{AB}\Omega_A^a\Omega_B^b + \frac{1}{2} I_{ab}^{BB}\Omega_B^a\Omega_B^b \,, \end{aligned} \tag{165}$$

and total moments of inertia acquire due to their sum, e.g. $I_{ab}^{AA} = I_{ab,\mathrm{val}}^{AA} + I_{ab,0}^{AA}$. where E_{static} is the self-consistent classical soliton energy and $D_{ab}(A) = \frac{1}{2}\mathrm{Tr}(\lambda_a A \lambda_b A^\dagger)$ is a $SU(3)$ Wigner rotation matrix.

In the evaluations of the moments of inertia, the numerical difficulties may arise. Due to the difference of the boundary conditions between the initial and the final states of the matrix elements one may obtain the spurious nonzero values of the moments of inertia in the absence of background pion fields. Similar problem occurs in the $\langle n|J_a|m\rangle$ for all values of a and $\langle n|\lambda_a|m\rangle$ for $a \geq 4$ with our harmonic oscillator basis. To avoid the problem, we employ the replacement of Eq. (112) for $\langle n|J_a|m\rangle$ [15]. For the matrix element of λ_a, similar replacement induces an additional spurious term proportional to the mass difference. However, within our perturbative treatment in the mass difference, this procedure is justified.

With the standard canonical quantization formula for the collective coordinates we obtain the following quantization prescriptions

$$R_a = \begin{cases} -\sum_j (I_{aj}^{AA}\Omega_A^j - I_{aj}^{AB}\Omega_B^j - K_{aj}^A D_{8j}), & a = 1,2,3, \\ -\sum_b (I_{ab}^{AA}\Omega_A^b - K_{ab}^A D_{8b}), & a = 4,5,6,7, \\ \frac{\sqrt{3}}{2}B, & a = 8, \end{cases} \tag{166}$$

and

$$K_i = -\sum_j (I_{ij}^{BB}\Omega_B^j - I_{ij}^{BA}\Omega_A^j + K_{ij}^B D_{8j}), \quad i = 1,2,3, \tag{167}$$

where R_a is the right isospin generator of $SU(3)$, and K_i represents the generator of the spatial rotation.

Due to the symmetry of the soliton, only the following elements of the moments of inertia survive:

$$\begin{aligned} & I_{11}^{AA} = I_{22}^{AA}, I_{11}^{BB} = I_{22}^{BB}, \\ & I_{33}^{BB} = m^2 I_{33}^{AA}, I_{33}^{AB} = I_{33}^{BA} = -mI_{33}^{AA}, \\ & I_{44}^{AA} = I_{55}^{AA} = I_{66}^{AA} = I_{77}^{AA}, \\ & K_{11}^{AA} = K_{22}^{AA}, \\ & K_{33}^{BB} = -mK_{33}^{AA}, \\ & K_{44}^{AA} = K_{55}^{AA} = K_{66}^{AA} = K_{77}^{AA}. \end{aligned} \tag{168}$$

Thus the hamiltonian becomes

$$\begin{aligned} H &= E_{\mathrm{static}} + H_0 + H_1, \\ H_0 &= \frac{1}{2}\frac{1}{I_{44}^{AA}}\sum_{a=1}^{7} R_a^2 + \frac{1}{2}\left(\frac{1}{I_{11}^{AA}} - \frac{1}{I_{44}^{AA}}\right)\sum_{i=1}^{3} R_i^2 \\ &+ \frac{1}{2}\frac{1}{I_{11}^{BB}}\sum_{i=1}^{3} K_i^2 + \frac{1}{2}\left(\frac{1}{I_{33}^{AA}} - \frac{1}{I_{11}^{AA}} - \frac{m^2}{I_{11}^{BB}}\right)R_3^2. \end{aligned} \tag{169}$$

In the evaluation of the H, we adopt a simple perturbative treatment with the mass difference Δm [59]. Up to first order of the Δm, the H_1 is written as

$$\begin{aligned} H_1 &= \frac{1}{2}\gamma(1 - D_{88}) - \frac{K^A_{11}}{I^{AA}_{11}}(D_{81}R_1 + D_{82}R_2) \\ &- \frac{K^A_{33}}{I^{AA}_{33}}D_{83}R_3 - \frac{K^A_{44}}{I^{AA}_{44}}\sum_{k=4}^{7} D_{8k}R_k . \end{aligned} \tag{170}$$

The hamiltonian in Eq. (169) is diagonalized by using following collective wave functions of the nonperturbative part of the hamiltonian H_0

$$\begin{aligned} &\Psi^{(n)}_{YII_3,Y'NN_3,JJ_3}(A,B) = \sqrt{\dim(n)}(-1)^{\frac{Y'}{2}+N_3} \\ &\times D^{(n)*}_{YII_3,Y'NN_3}(A) D^{J*}_{J_3,-mN_3}(B) . \end{aligned} \tag{171}$$

With these bases, the matrix element reduces to the integral of the three Wigner matrices which can be easily evaluated from the $SU(3)$ Clebsch-Gordan coefficients [59, 63]. The actual computations of the Clebsch-Gordan coefficients can be performed by using the numerical algorithm in Ref. [64].

Table 6: The various moments of inertia, with $m_s = 149$ MeV.

	Valence	Vacuum	Total
$I^{AA}_{11}[\text{GeV}^{-1}]$	7.49	4.05	11.54
$I^{BB}_{11}[\text{GeV}^{-1}]$	11.19	5.46	16.65
$I^{AA}_{33}[\text{GeV}^{-1}]$	4.36	1.61	5.97
$I^{AA}_{44}[\text{GeV}^{-1}]$	1.64	1.26	2.90
K^A_{11}	0.285	1.38×10^{-4}	0.285
K^A_{33}	0.297	1.60×10^{-3}	0.298
K^A_{44}	0.255	-1.05×10^{-3}	0.254
$\gamma[\text{MeV}]$	292.0	1098.8	1390.8

Our numerical calculations were performed with the constituent mass $M = 400$ MeV. For a diagonalization problem of the Dirac hamiltonian, we used the deformed harmonic oscillator basis [65] which is described in detail in Appendix A. The self-consistent classical mass was obtained as $E_{\text{static}} = 2406$ MeV which differs from the result of $SU(2)$ sector. The difference arises due to the specific choice of the cutoff scheme in Eqs. (137)-(139). The values of various moments of inertia are listed in Table 6. The quantized states are coupled to the multiplets $\{\overline{10}\}$, $\{27\}$, $\{35\}$, $\{28\}$ corresponding to $(p, q) = (0, 3), (2, 2), (4, 1), (6, 0)$ respectively. In Table 7, we show all the mass of the dibaryon states for the multiplets. The energy levels of the dibaryon states belonging to each multiplets are shown in Fig. 11.

In a pioneering work of the $SU(3)$ collective quantization of the chiral soliton [66], it is pointed out that because of the constraint $Y_R = 2$ which arises from the trivial embedding,

Table 7: Absolute mass of the dibaryon (in MeV), with $m_s = 149$ MeV.

Multiplet	$(S\ I\ J)$	Mass	Multiplet	$(S\ I\ J)$	Mass
$\overline{10}$	(0 0 1)	3255	35	(0 2 1)	3610
				(-1 $\frac{5}{2}$ 1)	3965
	(-1 $\frac{1}{2}$ 1)	3467		(-1 $\frac{3}{2}$ 1)	3727
				(-2 2 1)	4034
	(-2 1 1)	3679		(-2 1 1)	3844
				(-3 $\frac{3}{2}$ 1)	4103
	(-3 $\frac{3}{2}$ 1)	3891		(-3 $\frac{1}{2}$ 1)	3960
				(-4 1 1)	4172
				(-4 0 1)	4077
27	(0 1 0)	3309		(-5 $\frac{1}{2}$ 1)	4241
	(-1 $\frac{3}{2}$ 0)	3573			
	(-1 $\frac{1}{2}$ 0)	3453	28	(0 3 0)	3969
	(-2 2 0)	3841		(-1 $\frac{5}{2}$ 0)	4054
	(-2 1 0)	3678		(-2 2 0)	4139
	(-2 0 0)	3597		(-3 $\frac{3}{2}$ 0)	4224
	(-3 $\frac{3}{2}$ 0)	3904		(-4 1 0)	4309
	(-3 $\frac{1}{2}$ 0)	3781		(-5 $\frac{1}{2}$ 0)	4393
	(-4 1 0)	3966		(-6 0 0)	4478

some states in the constituent quark model are not allowed in the soliton solution. Hence the state we obtained is not the lowest state in the $S = -2$ sector and the configuration of H-dibaryon may have not be an axially symmetric. On the other hand, $(I, J) = (0, 0)$ channel in the $S = -6$ sector, corresponding to di-Omega $\Omega\Omega$ could have a rather deeper bound. According to the data of the $B = 1$ hedgehog analysis (in Ref. [59]), we expect the binding energy about $\sim$ 200 MeV. This state is rather promising as a candidate of the axially symmetric dibaryon.

In the perturbative treatment of Eqs. (165),(169) we retained only linear terms for the mass difference Δm and used $SU(3)$ symmetric wave functions in Eq. (171) for the ground state. For the $B = 1$ hedgehog case, such leading order approximation induces discrepancy with experiment [59]. But it is possible to minimize it by incorporating appropriate higher-order effects in the Δm [67]. In our estimation of moments of inertia, in order to remove the spurious contributions, we employ ad hoc approximation for the matrix elements and it is justified only if we confine our calculations up to first order. We expect an extension of our scheme to the second order in Δm is also feasible.

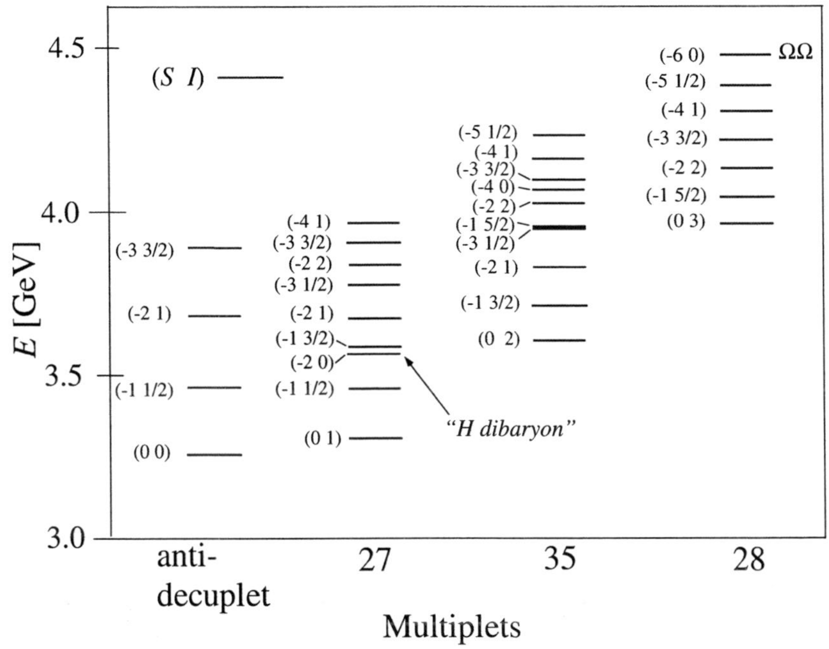

Figure 11: The dibaryon spectra [15].

6 Concluding Remarks

This article reports the theoretical framework and the numerical results of the chiral quark soliton model for higher baryon number solutions. The model was inspired by the instanton liquid model of the QCD vacuum and thus incorporates the basic features of QCD, e.g. the chiral symmetry and its breakdown accompanied by the appearance of the Goldstone bosons. This model provides correct observable as a nucleon including mass, electromagnetic value, spin carried by quarks, parton distributions and octet, decuplet $SU(3)$ baryon spectra. For $B > 1$, different topological configurations from $B = 1$ are needed to obtain the minimal energy solutions. For $B = 2$, we employed the axially symmetric configuration which produces the minimal energy configuration in the Skyrme model. We also investigated the $B = 3, 4, 5$ solitons with axial symmetry. The solution exhibits doubly degenerate bound spectra of the one-quark orbits. This relatively large degeneracy confirms that the solutions are stable local minima. For $B > 2$, the Skyrme model predicts that the solutions have only discrete symmetries. According to the prediction, we studied the CQSM with the chiral fields of such platonic symmetries. The discrete crystal-like symmetries exhibit much complicated structure and the study of such configurations is rather formidable task. However the analysis becomes much simpler when we adopt the rational map ansatz to the chiral fields since with this ansatz the chiral fields are separable in polar coordinates and radial coordinate, which makes the numerical technique developed for $B = 1$ applicable to find solutions with higher B. We showed that the baryon densities inherit the same discrete symmetries as the chiral fields and obtained various degenerate bound spectra of the valence quarks depending on the background of chiral field configurations. Evaluating the radial component of the baryon density, shell-like structure of the valence quark spectra was observed. The group theory should predict these level structures resulting from the symmetry of the background potential. In fact the degeneracy of the valence spectra are determined by the winding number of the chiral fields as well as the shape deformation (symmetry) of solitons. The four-fold degeneracy of the lowest states may be ascribed to the chiral symmetry $SU(2)_L \times SU(2)_R$ of the hamiltonian. To get better understanding of the relation between the quark level structure and the winding number or the shape deformation, further analysis will be worth to be done in future.

Upon quantization, we computed zero-mode rotational corrections to the classical energy. The study of the Finkelstein-Rubinstein constraints indicates that the axially symmetric solution with even B has the same quantum number as the physically observed nuclei. Some of the states may be observed in experiments. For example, in the $B = 2$, we obtained $I = 0, J = 1$ ($^3S_1$: deuteron) and $I = 1, J = 0$ ($^1S_0$) solutions. The energy of $^3S_1$ is lower than the $^1S_0$. The order is in agreement with the experimental observations. For $B = 4$, the quantum number of the ground state $I = 0, J = 0$ coincide with the observation. For odd B, the constraint of C_2 in Eq. (126) seems to assure the validity of the ansatz. Indeed, it provides the ground state as $I = J = 1/2$ for $B = 3$ and as $I = 1/2, J = 3/2$ for $B = 5$ due to their shell-like structure, which coincide with physical observations. However, unfortunately the constraint in Eq. (124) forbid such states. Consequently, the axially symmetric solitons with odd B can appear only as excited states.

For an $SU(3)$ extension of the model, we adopted the collective quantization scheme with a trivial embedding form for the chiral fields. In order to estimate the effects of the quark mass difference, we performed the naive perturbative method in terms of the mass difference. We obtained the dibaryonic spectrum coupled to the multiplets of $\{\overline{10}\}$, $\{27\}$, $\{35\}$, $\{28\}$. The state we obtained is not the lowest state in the $S = -2$ sector. The configuration of H-dibaryon may have not be an axially symmetric. On the other hand, $(I, J) = (0, 0)$ channel in the $S = -6$ sector, corresponding to di-Omega $\Omega\Omega$ may have a rather deeper bound. This state is promising as a candidate of the axially symmetric dibaryon.

In our analysis, all obtained states seem to be deep bound states. Consideration of the Casimir effect is, however, necessary to determine which states are stable. For the $B = 1$ skyrmion, the Casimir energies of the rotational and the translational zero modes were estimated by various authors. Their predictions for the total mass are around $-0.5 \sim -1.3$ GeV [41, 42, 43, 44, 45]. For the $B = 2$ torus, the Casimir energies have not been estimated yet. The thorough analysis of the Casimir effects is desired in order to examine the stability.

A Numerical Basis

A.1 Kahana-Ripka Basis

The numerical method widely used in this model is based on the expansion of the Dirac spinor in terms of an appropriate orthogonal basis. The Kahana-Ripka basis [28] which was originally constructed for diagonalizing the hamiltonian with the chiral fields of $B = 1$ hedgehog ansatz is a plane-wave finite basis. The basis is discretized by imposing an appropriate boundary condition on the radial wave functions at the radius $r_{\rm max}$ chosen to be sufficiently larger than the soliton size. The basis is then made finite by including only those states with the momentum k as $k < k_{\rm max}$. The results should be, however, independent on $r_{\rm max}$ and $k_{\rm max}$.

The hamiltonian with hedgehog ansatz commutes with the parity and the grandspin operator given by

$$\boldsymbol{K} = \boldsymbol{j} + \boldsymbol{\tau}/2 = \boldsymbol{l} + \boldsymbol{\sigma}/2 + \boldsymbol{\tau}/2,$$

where $\boldsymbol{j}, \boldsymbol{l}$ are respectively total angular momentum and orbital angular momentum. Accordingly, the angular basis can be written as

$$|(lj)KM\rangle = \sum_{j_3\tau_3} C^{KM}_{jj_3\frac{1}{2}\tau_3}\Big(\sum_{m\sigma_3} C^{jj_3}_{lm\frac{1}{2}\sigma_3}|lm\rangle|\frac{1}{2}\sigma_3\rangle\Big)|\frac{1}{2}\tau_3\rangle\,. \tag{172}$$

With this angular basis, the normalized eigenstates of the free hamiltonian in a spherical box with radius $r_{\rm max}$ can be constructed as follows:

$$\begin{aligned}
u^{(1)}_{KM} &= N_k\begin{pmatrix} ij_K(kr)|(KK+\frac{1}{2})KM\rangle \\ \Delta_k j_{K+1}(kr)|(K+1K+\frac{1}{2})KM\rangle\end{pmatrix},\\
u^{(2)}_{KM} &= N_k\begin{pmatrix} ij_K(kr)|(KK-\frac{1}{2})KM\rangle \\ -\Delta_k j_{K-1}(kr)|(K-1K-\frac{1}{2})KM\rangle\end{pmatrix},\\
u^{(3)}_{KM} &= N_k\begin{pmatrix} i\Delta_k j_K(kr)|(KK+\frac{1}{2})KM\rangle \\ -j_{K+1}(kr)|(K+1K+\frac{1}{2})KM\rangle\end{pmatrix},\\
u^{(4)}_{KM} &= N_k\begin{pmatrix} i\Delta_k j_K(kr)|(KK-\frac{1}{2})KM\rangle \\ j_{K-1}(kr)|(K-1K-\frac{1}{2})KM\rangle\end{pmatrix},\\
v^{(1)}_{KM} &= N_k\begin{pmatrix} ij_{K+1}(kr)|(K+1K+\frac{1}{2})KM\rangle \\ -\Delta_k j_K(kr)|(KK+\frac{1}{2})KM\rangle\end{pmatrix},\\
v^{(2)}_{KM} &= N_k\begin{pmatrix} ij_{K-1}(kr)|(K-1K-\frac{1}{2})KM\rangle \\ \Delta_k j_K(kr)|(KK-\frac{1}{2})KM\rangle\end{pmatrix},\\
v^{(3)}_{KM} &= N_k\begin{pmatrix} i\Delta_k j_{K+1}(kr)|(K+1K+\frac{1}{2})KM\rangle \\ j_K(kr)|(KK+\frac{1}{2})KM\rangle\end{pmatrix},\\
v^{(4)}_{KM} &= N_k\begin{pmatrix} i\Delta_k j_{K-1}(kr)|(K-1K-\frac{1}{2})KM\rangle \\ -j_K(kr)|(KK-\frac{1}{2})KM\rangle\end{pmatrix},
\end{aligned} \tag{173}$$

with

$$N_k = \left[\frac{1}{2}r^3_{\rm max}\Big(j_{K+1}(kr_{\rm max})\Big)^2\right]^{-1/2} \tag{174}$$

and $\Delta_k = k/(E_k + M)$.

The momenta are discretized by the boundary condition $j_K(k_i r_{\rm max}) = 0$. The u, v correspond to the *"natural"* and *"unnatural"* components of the basis which stand for parity $(-1)^K$ and $(-1)^{K+1}$ respectively.

Let us construct the trial function using the Kahana-Ripka basis to solve the eigenequations in Eq. (7),

$$\phi_\mu(\boldsymbol{x}) = \lim_{K_{\max}\to\infty} \sum_{K=0}^{K_{\max}} \sum_{M=-K}^{K} \sum_{j=1}^{4} [\alpha^{(j)}_{KM,\mu} u^{(j)}_{KM}(r,\theta,\varphi) + \beta^{(j)}_{KM,\mu} v^{(j)}_{KM}(r,\theta,\varphi)]. \tag{175}$$

A.2 Deformed Oscillator Basis

Let us show the numerical analysis of the eigen equations in detail. To solve the eigenequation of the form,

$$[-i\boldsymbol{\alpha}\cdot\nabla + \beta M(\cos F(\rho,z) + i\gamma_5 \boldsymbol{\tau}\cdot\hat{\boldsymbol{n}}_R \sin F(\rho,z)]\phi_\mu(\boldsymbol{x}) = E_\mu \phi_\mu(\boldsymbol{x}), \tag{176}$$

we introduce the deformed harmonic oscillator spinor basis which was originally constructed by Gambhir *et al.* in the relativistic mean field theory for deformed nuclei [65]. The upper and lower components of the Dirac spinors are expanded separately by the basis as

$$\phi_\mu(\boldsymbol{x}) = \begin{pmatrix} f_\mu(\boldsymbol{x}) \\ i g_\mu(\boldsymbol{x}) \end{pmatrix} = \begin{pmatrix} \sum_a f_{\mu a} \boldsymbol{\Phi}_a(\boldsymbol{x}, s) \\ i \sum_{\tilde{a}} g_{\mu\tilde{a}} \boldsymbol{\Phi}_{\tilde{a}}(\boldsymbol{x}, s) \end{pmatrix} \chi^I_{m_\tau} \tag{177}$$

where $\boldsymbol{\Phi}_a(\boldsymbol{x}, s, \tau)$ are the eigefunctions of a deformed harmonic oscillator potential

$$V_{osc}(\rho, z) = \frac{1}{2} M\omega_\rho^2 \rho^2 + \frac{1}{2} M\omega_z^2 z^2, \tag{178}$$

and defined by

$$\boldsymbol{\Phi}_a(\boldsymbol{x}, m_s) = \frac{1}{\sqrt{2\pi}} \phi^{|\omega|}_{n_r}(\rho) \phi_{n_z}(z) e^{i\omega\varphi} \chi^S_{m_s} \tag{179}$$

with

$$\begin{aligned} \phi^{|\omega|}_{n_r}(\rho) &= N^{|\omega|}_{n_r} (\sqrt{\alpha_\rho}\rho)^{|\omega|} e^{-\frac{1}{2}\alpha_\rho \rho^2} L^{|\omega|}_{n_r}(\alpha_\rho \rho^2) \\ & n_r = 0, 1, 2, \cdots, N_{\text{rmax}} \\ \phi_{n_z}(z) &= N_{n_z} e^{-\frac{1}{2}\alpha_z z^2} H_{n_z}(\sqrt{\alpha_z} z) \\ & n_z = 1, 3, \cdots, 2N_{\text{zmax}} + 1 \ \text{ or } \ 0, 2, \cdots, 2N_{\text{zmax}}, \end{aligned}$$

and

$$\chi_+ = \begin{pmatrix} 1 \\ 0 \end{pmatrix}, \quad \chi_- = \begin{pmatrix} 0 \\ 1 \end{pmatrix} \tag{180}$$

depending on if the eigenvalues of the third components of the spin m_s (isospin m_τ) takes $+1$ or -1. The functions, $L^{|m|}_{n_r}$ and H_{n_z}, are the associated Laguerre polynomials and the Hermite polynomials with the normalization constants

$$N^{|\omega|}_{n_r} = \sqrt{\frac{2\alpha_\rho n_r!}{(n_r + |\omega|)!}}, \quad N_{n_z} = \frac{1}{\sqrt{2^{n_z} n_z! \sqrt{\frac{\pi}{\alpha_z}}}}. \tag{181}$$

These polynomials can be calculated by following recursion relations

$$x\frac{d}{dx}L_n^m(x) = nL_n^m(x) - (n+m)L_{n-1}^m(x) \tag{182}$$

$$L_n^{m-1}(x) = L_n^m(x) - L_{n-1}^m(x) \tag{183}$$

and

$$H_{n+1}(x) - 2xH_n(x) + 2n_zH_{n-1}(x) = 0 \tag{184}$$

$$\frac{d}{dx}H_n(x) = 2nH_{n-1}(x) \tag{185}$$

where constants α_ρ and α_z can be expressed by the oscillator frequencies as

$$\alpha_\rho = \frac{\mathcal{M}\omega_\rho}{\hbar},\ \alpha_z = \frac{\mathcal{M}\omega_z}{\hbar} \tag{186}$$

which are free parameters chosen optimally. The N_{rmax} and N_{zmax} are increased until convergence is attained. The parity transformation rule of Φ_α is given by

$$\Phi_\alpha(\rho, \varphi+\pi, -z; s, t) = (-1)^{\omega+n_z}\Phi_\alpha(\rho, \varphi, z; s, t) \tag{187}$$

where

$$H_{n_z}(-\sqrt{\alpha_z}z) = (-1)^{n_z}H_{n_z}(\sqrt{\alpha_z}z), \tag{188}$$

has been used. The parity is $+$ for $\omega + n_z =$ odd, and $-$ for $\omega + n_z =$ even.

There are two sets of the complete basis for each parity. One is the natural basis with $K_3^P = 0^+, 1^-, 2^+, \cdots$, for odd B and $K_3^P = \frac{1}{2}^+, \frac{3}{2}^-, \frac{5}{2}^+, \cdots$ for even B. Another is the unnatural basis with $K_3^P = 0^-, 1^+, 2^-, \cdots$, for odd B and $K_3^P = \frac{1}{2}^-, \frac{3}{2}^+, \frac{5}{2}^-, \cdots$ for even B The natural basis is given by

$$\begin{aligned}\phi_\mu^n(\boldsymbol{x}) =& \\ &\begin{pmatrix} \sum\limits_{\alpha(0)} f_{\alpha(0),\mu}\boldsymbol{\Phi}_{\alpha(0)}(\boldsymbol{x},\uparrow s) + \sum\limits_{\alpha(1)} f_{\alpha(1),\mu}\boldsymbol{\Phi}_{\alpha(1)}(\boldsymbol{x},\downarrow s) \\ i\sum\limits_{\beta(0)} g_{\beta(0),\mu}\boldsymbol{\Phi}_{\beta(0)}(\boldsymbol{x},\uparrow s) + i\sum\limits_{\beta(1)} g_{\beta(1),\mu}\boldsymbol{\Phi}_{\beta(1)}(\boldsymbol{x},\downarrow s) \end{pmatrix}\chi_u^I \\ +&\begin{pmatrix} \sum\limits_{\alpha(2)} f_{\alpha(2),\mu}\boldsymbol{\Phi}_{\alpha(2)}(\boldsymbol{x},\uparrow s) + \sum\limits_{\alpha(3)} f_{\alpha(3),\mu}\boldsymbol{\Phi}_{\alpha(3)}(\boldsymbol{x},\downarrow s) \\ i\sum\limits_{\beta(2)} g_{\beta(2),\mu}\boldsymbol{\Phi}_{\beta(2)}(\boldsymbol{x},\uparrow s) + i\sum\limits_{\beta(3)} g_{\beta(3),\mu}\boldsymbol{\Phi}_{\beta(3)}(\boldsymbol{x},\downarrow s) \end{pmatrix}\chi_d^I\end{aligned} \tag{189}$$

where

$$\begin{aligned}\alpha(0) &= \{n_r, n_z : \text{odd},\ \omega_0 \equiv K_3 - 1/2 - m_{\text{w}}/2\} \\ \alpha(1) &= \{n_r, n_z : \text{even}, \omega_1 \equiv K_3 + 1/2 - m_{\text{w}}/2\} \\ \alpha(2) &= \{n_r, n_z : \text{even}, \omega_2 \equiv K_3 - 1/2 + m_{\text{w}}/2\} \\ \alpha(3) &= \{n_r, n_z : \text{odd},\ \omega_3 \equiv K_3 + 1/2 + m_{\text{w}}/2\}\end{aligned}$$

and

$$\begin{aligned}\beta(0) &= \{n_r, n_z : \text{even}, \omega_0 \equiv K_3 - 1/2 - m_{\rm w}/2\}\\ \beta(1) &= \{n_r, n_z : \text{odd}, \ \omega_1 \equiv K_3 + 1/2 - m_{\rm w}/2\}\\ \beta(2) &= \{n_r, n_z : \text{odd}, \ \omega_2 \equiv K_3 - 1/2 + m_{\rm w}/2\}\\ \beta(3) &= \{n_r, n_z : \text{even}, \omega_3 \equiv K_3 + 1/2 + m_{\rm w}/2\}.\end{aligned}$$

The unnatural basis $\phi_\mu^{(u)}$ is given by replacing, $\alpha \leftrightarrow \beta$ in Eq. (189). By using the natural and unnatural basis, the eigenvalue problem in Eq. (176) can be reduced to a symmetric matrix diagonalization problem.

Let us calculate the matrix elements of the hamiltonian below. For the kinetic term $\boldsymbol{\alpha}\cdot\boldsymbol{p}$, we have

$$\begin{aligned}&\langle\boldsymbol{\Phi}_{\alpha(0)}|\boldsymbol{\sigma}\cdot\boldsymbol{p}|i\boldsymbol{\Phi}_{\beta'(0)}\rangle\\ &= \frac{1}{2\pi}\int d^3x\phi_{n_r}^{|\omega_0|}(\rho)\phi_{n_z}(z)e^{-i\omega_0\varphi}\Big(\frac{\partial}{\partial z}\Big)\phi_{n'_r}^{|\omega'_0|}(\rho)\phi_{n'_z}(z)e^{i\omega'_0\varphi}\\ &= \delta_{n_rn'_r}(N_{n_z}N_{n'_z}\sqrt{\alpha_z}n'_z\frac{1}{N_{n_z}^2}\delta_{n_zn'_z-1} - \frac{1}{2}N_{n_z}N_{n'_z}\sqrt{\alpha_z}\frac{1}{N_{n_z}^2}\delta_{n_zn'_z+1})\\ &= \begin{cases}\delta_{\omega_0\omega'_0}\delta_{n_rn'_r}\frac{N'_{n_z}}{N_{n_z}}\sqrt{\alpha_z}n'_z\delta_{n_zn'_z-1}\\ \delta_{\omega_0\omega'_0}\delta_{n_rn'_r}(-\frac{1}{2})\frac{N_{n'_z}}{N_{n_z}}\sqrt{\alpha_z}\delta_{n_zn'_z+1},\end{cases}\end{aligned} \quad (190)$$

$$\begin{aligned}&\langle\boldsymbol{\Phi}_{\alpha(0)}|\boldsymbol{\sigma}\cdot\boldsymbol{p}|i\boldsymbol{\Phi}_{\beta'(1)}\rangle\\ &= \frac{1}{2\pi}\int d^3x\phi_{n_r}^{|\omega_0|}(\rho)\phi_{n_z}(z)e^{-i\omega_0\varphi}e^{-i\varphi}\Big(\frac{\partial}{\partial\rho} - \frac{i}{\rho}\frac{\partial}{\partial\varphi}\Big)\phi_{n'_r}^{|\omega'_1|}(\rho)\phi_{n'_z}(z)e^{i\omega'_1\varphi}\\ &= \begin{cases}\delta_{n_zn'_z}\sqrt{\alpha_r}(\sqrt{n_r+\omega_0+1}\delta_{n_rn'_r} + \sqrt{n_r}\delta_{n_r-1n'_r})\\ \qquad(\omega_0 \geq 0 : \omega'_1 = \omega_0 + 1 > 0)\\ -\delta_{n_zn'_z}\sqrt{\alpha_r}(\sqrt{n_r - \omega_0}\delta_{n_rn'_r} + \sqrt{n_r+1}\delta_{n_rn'_r-1}).\\ \qquad(\omega_0 < 0 : \omega'_1 = \omega_0 + 1 \leq 0)\end{cases}\end{aligned} \quad (191)$$

In the natural basis, quantum numbers (n_z, n'_z) takes values $(1,2),(3,4),\cdots$ for the upper part and $(1,0),(3,2),\cdots$ for the lower part. In the unnatural basis, $(n_z, n'_z) = (0,1),(2,3),\cdots$ for the upper part and $(n_z, n'_z) = (2,1),(3,2),\cdots$ for the lower part.

For the potential term $\beta M(\cos F(\rho,z) + i\gamma_5\boldsymbol{\tau}\cdot\hat{\boldsymbol{n}}_R\sin F(\rho,z))$, we have

$$\begin{aligned}&\langle\boldsymbol{\Phi}_{\alpha(0)}\chi_u^I|M\cos F(\rho,z)|\boldsymbol{\Phi}_{\alpha'(0)}\chi_u^I\rangle\\ &= \int \rho d\rho dz M\cos F(\rho,z)\phi_{n_r}^{|\omega_0|}(\rho)\phi_{n_z}(z)\phi_{n'_r}^{|\omega_0|}(\rho)\phi_{n'_z}(z),\end{aligned} \quad (192)$$

$$\begin{aligned}&\langle\boldsymbol{\Phi}_{\alpha(0)}\chi_u^I|Mi\boldsymbol{\tau}\cdot\hat{\boldsymbol{n}}_R\sin F(\rho,z)|i\boldsymbol{\Phi}_{\beta'(0)}\chi_u^I\rangle\\ &= -\int \rho d\rho dz M\cos\Theta(\rho,z)\sin F(\rho,z)\phi_{n_r}^{|\omega_0|}(\rho)\phi_{n_z}(z)\phi_{n'_r}^{|\omega'_0|}(\rho)\phi_{n'_z}(z),\end{aligned} \quad (193)$$

$$\begin{aligned}&\langle\boldsymbol{\Phi}_{\alpha(0)}\chi_u^I|Mi\boldsymbol{\tau}\cdot\hat{\boldsymbol{n}}_R\sin F(\rho,z)|\boldsymbol{\Phi}_{\beta'(2)}\chi_d^I\rangle\\ &= -\int \rho d\rho dz M\sin\Theta(\rho,z)\sin F(\rho,z)\phi_{n_r}^{|\omega_0|}(\rho)\phi_{n_z}(z)\phi_{n'_r}^{|\omega_2|}(\rho)\phi_{n'_z}(z).\end{aligned} \quad (194)$$

Other elements can be calculated in the same manner.

To estimate the moments of inertia, *e.g.*, $\langle n|\lambda_a|m\rangle$ let us introduce the basis for the strange direction [61]. The hamiltonian for the strange quark commutes with the total angular momentum

$$J_3 = L_3 + \frac{1}{2}\sigma_3 \tag{195}$$

because of the trivial construction of the $SU(3)$ chiral fields in Eq. (141). As a result, the deformed basis has common form with the $SU(2)$ (see Eqs. (179)-(189)), except in the quantum number in Eq. (189) as

$$\begin{aligned}\alpha(0) &= \{n_r, n_z : \text{even}, \omega_0 \equiv J_3 - 1/2\}\\ \alpha(1) &= \{n_r, n_z : \text{odd}, \ \omega_1 \equiv J_3 + 1/2\}\end{aligned}$$

and

$$\begin{aligned}\beta(0) &= \{n_r, n_z : \text{odd}, \ \omega_0 \equiv J_3 - 1/2\}\\ \beta(1) &= \{n_r, n_z : \text{even}, \omega_1 \equiv J_3 + 1/2\}\,.\end{aligned}$$

The unnatural basis $\phi_\mu^{(u)}$ is given by replacing, $\alpha \leftrightarrow \beta$ in Eq. (189).

B Operation $\hat{R}$ to the Numerical Basis

In this appendix, we present the construction of an operator $\hat{R}$ to the Kahana-Ripka basis in detail. Since it operates on the spherical harmonics, let us write the spherical harmonics in terms of the complex variables z and their conjugate $\bar{z}$ as follows :

$$z = \tan\frac{\theta}{2}e^{i\varphi}\,, \tag{196}$$

which can be related to the usual polar coordinates by

$$\cos\theta = \frac{1-|z|^2}{1+|z|^2}\,,\quad \exp(i\varphi) = \pm\sqrt{\frac{z}{\bar{z}}} \tag{197}$$

According to the definition of the spherical harmonics [68], we obtain (up to $l \leq 3$)

$$Y_{11} = -\frac{1}{2}\sqrt{\frac{3}{2\pi}}\frac{2z}{1+|z|^2}, \quad Y_{10} = \frac{1}{2}\sqrt{\frac{3}{\pi}}\frac{1-|z|^2}{1+|z|^2}, \quad Y_{1-1} = \frac{1}{2}\sqrt{\frac{3}{2\pi}}\frac{2\bar{z}}{1+|z|^2}, \tag{198}$$

$$\begin{aligned} Y_{22} &= \frac{1}{4}\sqrt{\frac{3\cdot 5}{2\pi}}\Big(\frac{2z}{1+|z|^2}\Big)^2, \quad Y_{21} = -\frac{1}{2}\sqrt{\frac{3\cdot 5}{2\pi}}\frac{2z(1-|z|^2)}{(1+|z|^2)^2}, \\ Y_{20} &= \frac{1}{4}\sqrt{\frac{5}{\pi}}\Big(3\Big(\frac{1-|z|^2}{1+|z|^2}\Big)^2 - 1\Big), \\ Y_{2-1} &= \frac{1}{2}\sqrt{\frac{3\cdot 5}{2\pi}}\frac{2\bar{z}(1-|z|^2)}{(1+|z|^2)^2}, \quad Y_{2-2} = \frac{1}{4}\sqrt{\frac{3\cdot 5}{2\pi}}\Big(\frac{2\bar{z}}{1+|z|^2}\Big)^2, \end{aligned} \tag{199}$$

$$\begin{aligned} Y_{33} &= -\frac{1}{8}\sqrt{\frac{5\cdot 7}{\pi}}\Big(\frac{2z}{1+|z|^2}\Big)^3, \quad Y_{32} = \frac{1}{4}\sqrt{\frac{3\cdot 5\cdot 7}{2\pi}}\frac{4z^2(1-|z|^2)}{(1+|z|^2)^3}, \\ Y_{31} &= -\frac{1}{8}\sqrt{\frac{3\cdot 7}{\pi}}\Big(5\Big(\frac{1-|z|^2}{1+|z|^2}\Big)^2 - 1\Big)\frac{2z}{1+|z|^2}, \\ Y_{30} &= \frac{1}{4}\sqrt{\frac{7}{\pi}}\Big(5\Big(\frac{1-|z|^2}{1+|z|^2}\Big)^2 - 3\Big)\frac{1-|z|^2}{1+|z|^2}, \\ Y_{3-1} &= \frac{1}{8}\sqrt{\frac{3\cdot 7}{\pi}}\Big(5\Big(\frac{1-|z|^2}{1+|z|^2}\Big)^2 - 1\Big)\frac{2\bar{z}}{1+|z|^2}, \\ Y_{3-2} &= \frac{1}{4}\sqrt{\frac{3\cdot 5\cdot 7}{2\pi}}\frac{4\bar{z}^2(1-|z|^2)}{(1+|z|^2)^3}, \quad Y_{3-3} = \frac{1}{8}\sqrt{\frac{5\cdot 7}{\pi}}\Big(\frac{2\bar{z}}{1+|z|^2}\Big)^3. \end{aligned} \tag{200}$$

According to the (inverse) transformation of the $Z_2 \times Z_2$

$$z \to -\frac{1}{z} \quad \equiv \quad \theta \to \pi - \theta, \varphi \to \pi - \varphi\,, \tag{201}$$

the spherical harmonics are transformed via

$$Y_{lm}(\theta,\varphi) \to Y_{lm}(\pi-\theta,\pi-\varphi) = (-1)^{l-m} Y_{l-m}(\theta,\varphi)\,. \tag{202}$$

For the (inverse) tetrahedral transformation, we have

$$z \to i\,\frac{1-z}{1+z}, \tag{203}$$

Thus the transformation of the spherical harmonics is given by

$$\begin{aligned}
Y_{11} &\to -\frac{i}{2}Y_{11} - \frac{i}{\sqrt{2}}Y_{10} - \frac{i}{2}Y_{1-1}\,,\\
Y_{10} &\to -\frac{i}{\sqrt{2}}Y_{11} + \frac{i}{\sqrt{2}}Y_{1-1}\,,\\
Y_{1-1} &\to \frac{i}{2}Y_{11} + \frac{i}{\sqrt{2}}Y_{10} + \frac{i}{2}Y_{1-1}\,.
\end{aligned} \tag{204}$$

$$\begin{aligned}
Y_{22} &\to -\frac{1}{4}Y_{22} - \frac{1}{2}Y_{21} - \frac{1}{2}\sqrt{\frac{3}{2}}Y_{20} - \frac{1}{2}Y_{2-1} - \frac{1}{4}Y_{2-2}\,,\\
Y_{21} &\to \frac{i}{\sqrt{2}}Y_{22} + \frac{i}{\sqrt{2}}Y_{21} - \frac{i}{2}Y_{2-1} - \frac{i}{2}Y_{2-2}\,,\\
Y_{20} &\to \frac{1}{2}\sqrt{\frac{3}{2}}Y_{22} - \frac{1}{2}Y_{20} + \frac{1}{2}\sqrt{\frac{3}{2}}Y_{2-2}\,,\\
Y_{2-1} &\to -\frac{i}{\sqrt{2}}Y_{22} + \frac{i}{\sqrt{2}}Y_{21} - \frac{i}{2}Y_{2-1} + \frac{i}{2}Y_{2-2}\,,\\
Y_{2-2} &\to -\frac{1}{4}Y_{22} + \frac{1}{2}Y_{21} - \frac{1}{2}\sqrt{\frac{3}{2}}Y_{20}\,, + \frac{1}{2}Y_{2-1} - \frac{1}{4}Y_{2-2}\,.
\end{aligned} \tag{205}$$

$$\begin{aligned}
Y_{33} &\to \frac{i}{8}Y_{33} + \frac{\sqrt{6}i}{8}Y_{32} + \frac{\sqrt{15}i}{8}Y_{31} + \frac{\sqrt{5}i}{4}Y_{30} + \frac{\sqrt{15}i}{8}Y_{3-1} + \frac{\sqrt{6}i}{8}Y_{3-2} + \frac{i}{8}Y_{3-3}\,,\\
Y_{32} &\to \frac{1}{4}\sqrt{\frac{3}{2}}Y_{33} + \frac{1}{2}Y_{32} + \frac{1}{4}\sqrt{\frac{5}{2}}Y_{31} - \frac{1}{4}\sqrt{\frac{5}{2}}Y_{3-1} - \frac{1}{2}Y_{3-2} - \frac{1}{4}\sqrt{\frac{3}{2}}Y_{3-3}\,,\\
Y_{31} &\to -\frac{\sqrt{15}i}{8}Y_{33} - \frac{\sqrt{10}i}{8}Y_{32} + \frac{i}{8}Y_{31} + \frac{\sqrt{3}i}{4}Y_{30} + \frac{i}{8}Y_{3-1} - \frac{\sqrt{10}i}{8}Y_{3-2} - \frac{\sqrt{15}i}{8}Y_{3-3}\,,\\
Y_{30} &\to -\frac{\sqrt{5}}{4}Y_{33} + \frac{\sqrt{3}}{4}Y_{31} - \frac{\sqrt{3}}{4}Y_{3-1} + \frac{\sqrt{5}}{4}Y_{3-3}\,,\\
Y_{3-1} &\to \frac{\sqrt{15}i}{8}Y_{33} - \frac{\sqrt{10}i}{8}Y_{32} - \frac{i}{8}Y_{31} + \frac{\sqrt{3}i}{4}Y_{30} - \frac{i}{8}Y_{3-1} - \frac{\sqrt{10}i}{8}Y_{3-2} + \frac{\sqrt{15}i}{8}Y_{3-3}\,,\\
Y_{3-2} &\to \frac{1}{4}\sqrt{\frac{3}{2}}Y_{33} - \frac{1}{2}Y_{32} + \frac{1}{4}\sqrt{\frac{5}{2}}Y_{31} - \frac{1}{4}\sqrt{\frac{5}{2}}Y_{3-1} + \frac{1}{2}Y_{3-2} - \frac{1}{4}\sqrt{\frac{3}{2}}Y_{3-3}\,,\\
Y_{3-3} &\to -\frac{i}{8}Y_{33} + \frac{\sqrt{6}i}{8}Y_{32} - \frac{\sqrt{15}i}{8}Y_{31} + \frac{\sqrt{5}i}{4}Y_{30} - \frac{\sqrt{15}i}{8}Y_{3-1} + \frac{\sqrt{6}i}{8}Y_{3-2} - \frac{i}{8}Y_{3-3}\,.
\end{aligned} \tag{206}$$

The operation $\hat{R}$ to the basis becomes

$$\hat{R}|(lj)KM\rangle \equiv \hat{K}\sum_{j_3\tau_3} C^{KM}_{jj_3\frac{1}{2}\tau_3}\Big(\sum_{m\sigma_3} C^{jj_3}_{lm\frac{1}{2}\sigma_3}|lm\rangle'|\frac{1}{2}\sigma_3\rangle\Big)|\frac{1}{2}\tau_3\rangle\,, \tag{207}$$

where $|lm\rangle'$ representa the transformations of the spherical harmonics in Eqs. (202) or (205)-(207). Evaluationg $\hat{R}$ for each symmetric operation, one can get the final results in Eqs. (93)-(97).

References

[1] S. J. Freedman *et al.*, *Phys. Rev. C* **48**, 1864 (1993);

[2] F. Gross and S. Luiti, *Phys. Rev. C* **45** 1374 (1992).

[3] W. Melnitchouk, A. W. Schreiber, and A. W. Thomas, *Phys. Rev. D* **49**, 1183 (1994).

[4] A. Yu. Umnikov, F. C. Khanna and L. P. Kaptari, *Phys. Rev. C* **53**, 377 (1996).

[5] C. E. Carlson and K. E. Lassila, *Phys. Rev. C* **51**, 364 (1995).

[6] D. Diakonov, V. Petrov and M. Polykov, *Z. Phys. A* **359**, 305 (1997).

[7] T. Nakano *et al.*, *Phys. Rev. Lett.* **91**, 012002 (2003).

[8] D. I. Diakonov, V. Yu. Petrov, and P. V. Pobylitsa, *Nucl. Phys. B* **306**, 809 (1988).

[9] H. Reinhardt and R. Wünsch , *Phys. Lett. B* **215**, 577 (1988).

[10] Th. Meissner, F. Grümmer, and K. Goeke, *Phys. Lett. B* **227**, 296 (1989).

[11] For detailed reviews of the model see: R. Alkofer, H. Reinhardt and H. Weigel, *Phys. Rept.* **265**, 139 (1996); Chr. V. Christov, A. Blotz, H.-C. Kim, P. Pobylitsa, T. Watabe, Th. Meissner, E. Ruiz Arriola, K. Goeke, *Prog. Part. Nucl. Phys.* **37**, 91 (1996).

[12] M. Wakamatsu and H. Yoshiki, *Nucl. Phys. A* **524**, 561 (1991).

[13] N. Sawado and S. Oryu, *Phys. Rev. C* **58**, R3046 (1998).

[14] N. Sawado, *Phys. Rev. C* **61**, 65206 (2000).

[15] N. Sawado, *Phys. Lett. B* **524**, 289 (2002).

[16] E. Braaten, S. Townsend and L. Carson, *Phys. Lett. B* **235**, 147 (1990).

[17] R.A. Battye and P.M. Sutcliffe, *Phys. Rev. Lett.* **79**,363 (1997).

[18] C. J. Houghton, N. S. Manton and P. M. Sutcliffe, *Nucl. Phys. B* **510**, 507 (1998).

[19] N. Sawado and N. Shiiki, *Phys. Rev. D* **66**, 011501 (2002).

[20] Nobuyuki Sawado and Noriko Shiiki, hep-ph/0402084.

[21] S. Komori, N. Sawado and N. Shiiki, *Ann. Phys.* **311**, 1 (2004).

[22] P. Irwin, *Phys. Rev. D* **61**, 114024 (2000).

[23] D. Y. Grigoriev, P. M. Sutcliffe and D. H. Tchrakian, *Phys. Lett. B* **540**, 146 (2002).

[24] R. L. Jaffe, *Phys. Rev. Lett.* **38**, 195 (1977); 38, 1617(E) (1977).

[25] A. Dhar, R. Shankar, S.R. Wadia, *Phys. Rev. D* **31**, 3256 (1985).

[26] D. Ebert, H. Reinhardt, *Nucl. Phys. B* **271**, 188,(1986).

[27] I. Aitchison, C. Fraser, E. Tudor and J. Zuk, *Phys. Lett. B* **165**, 162 (1985).

[28] S. Kahana, G. Ripka, and V. Soni, *Nucl. Phys. A* **415**, 351 (1984); S. Kahana and G. Ripka, *ibid*, A 429, 462 (1984).

[29] A. P. Balachandran and S. Vaidya, *Int. J. Mod. Phys. A* **14**, 445 (1999), hep-th/9803125.

[30] R. Rajaraman, *Solitons and Instantons* (North-Holland Physics Publishing, Amsterdam, 1987).

[31] H. Reinhardt, *Nucl. Phys. A* **503**, 825 (1989).

[32] J. Schwinger, *Phys. Rev.* **82**, 664 (1951).

[33] N. S. Manton, *Phys. Lett. B* **192,** 177 (1987).

[34] V. B. Kopeliovich and B. E. Stern, *Pis'ma Zh. Éksp. Teor. Fiz.* **45,** 165 (1987) [*JETP Lett.* **45**, 203 (1987)].

[35] J. Verbaarschot, *Phys. Lett. B* **195**, 235 (1987).

[36] H. Weigel, B. Schwesinger and G. Holzwarth, *Phys. Lett. B* **168**, 321 (1986).

[37] E. Braaten and L. Carson, *Phys. Rev. D* **38**, 3525 (1988).

[38] R. A. Battye and P. M. Sutcliffe, *Rev. Math. Phys.* **14**, 29, (2002).

[39] A. Bohr and B. Mottelson, *Nuclear structure, Vol. II* (World Scientific Publishing Co. Pte. Ltd, Singapore, 1998).

[40] N. S. Manton and B. M. A. G. Piette, *Prog. Math.* **201**, 469 (2001); hep-th/0008110.

[41] B. Moussallam and D. Kalafatis, *Phys. Lett. B* **272**, 196 (1991).

[42] B. Moussallam, *Ann. Phys.* **225**, 264 (1993).

[43] G. Holzwarth and H. Walliser, *Nucl. Phys. A* **587**, 721 (1995).

[44] H. Weigel, R. Alkofer and H. Reinhardt, *Nucl. Phys. A* **582**, 484 (1995).

[45] F. G. Scholtz, B. Schwesinger and H. B. Geyer, *Nucl. Phys. A* **561**, 542 (1993).

[46] N. S. Manton, *Phys. Lett. B* **110**, 54 (1982).

[47] N. S. Manton, *Nucl. Phys. B* **126**, 525 (1977).

[48] E. B. Bogomol'nyi, *Sov. J. Phys.* **24**, 97 (1977).

[49] N. Sawado, N. Shiiki and Y. Tanaka, *unpublished*.

[50] G. W. Gibbons and N. S. Manton, *Nucl. Phys. B* **274**, 183 (1986).

[51] Robert Leese, *Nucl. Phys. B* **344**, 33 (1990).

[52] J. Dudek, A. Goździ, N. Schunck, and M. Miśkiewicz, *Phys. Rev. Lett.* **88**, 252502 (2002).

[53] M. Hamermesh, *Group theory and its application to physical problems* (Dover Publications, Inc., New York, 1989).

[54] W. Koepf and P. Ring, *Nucl. Phys. A* **493**, 61 (1989).

[55] L. C. Biedenharn, Y. Dothan and M. Tarlini *Phys. Rev. D* **31,** 649 (1985).

[56] K. Goeke, A. Z. Górski, F. Grümmer, Th. Meissner, H. Reinhardt, and R. Wünsch, *Phys. Lett. B* **256**, 321 (1991).

[57] D. Finkelstein and J. Rubinstein, *J. Math. Phys.* **9**, 1762 (1968).

[58] S. Krusch, *Ann. Phys.* **304**, 103 (2003); hep-th/0210310.

[59] A. Blotz, D. Diakonov, K. Goeke, N. W. Park, V. Petrov and P. V. Pobylitsa, *Nucl. Phys. A* **555**, 765 (1993).

[60] M. Wakamatsu, *Phys. Rev. D* **54**, 2161 (1996).

[61] H. Weigel, R. Alkofer and H. Reinhardt, *Nucl. Phys. B* **387**, 638 (1992).

[62] V. B. Kopeliovich, B. Schwesinger and B. E. Stern, *Nucl. Phys. A* **549**, 485 (1992).

[63] N. Toyota, *Prog. Theor. Phys.* **77**, 688 (1987).

[64] T. A. Kaeding, H. T. Williams, *Comput. Phys. Commun.* **98**, 398 (1996).

[65] Y. K. Gambhir, P. Ring, A. Thimet, *Ann. Phys.* **198**, 132 (1990).

[66] J. Kunz and P. J. Mulders, *Phys. Lett. B* **215**, 449 (1988).

[67] A. Blotz, M. Praszalowicz and K. Goeke, *Phys. Rev. D* **53**, 485 (1996).

[68] Milton Abramowitz and Irene A. Stegun, *Handbook of Mathematical Functions* (Dover Publications, New York, 1968).

In: New Developments in Soliton Research
Editor: L.V. Chen pp. 53-85
ISBN 1-59554-561-8

Chapter 2

SOLITONS AND TOPOLOGICAL SOLUTIONS ON A FUZZY SPHERE

P. Valtancoli
Dipartimento di Fisica, Polo Scientifico Universitá di Firenze
and INFN, Sezione di Firenze (Italy)
Via G. Sansone 1, 50019 Sesto Fiorentino, Italy

Abstract

We review our recent construction of soliton and topological solutions of noncommutative gauge theory on a fuzzy sphere. To reach these goals we have employed several methods, i.e. the matrix model approach and the projective module one, introducing a new general link between them. In the case of search for solitons we have firstly introduced an ansatz into the general set of classical equations of motion of the matrix model, obtaining a general class of solutions. Between them we have selected those admitting a classical limit, which in turn require that a certain coupling constant λ of the model behaves as $\lambda = \frac{1}{2} + \frac{\lambda_0}{N}$ in the $N \to \infty$ limit. In this way we find classical solitons characterized by the angular momentum $l = 2$. More general type of ansatze are required to search for solitons with general angular momentum l. This technique is not enough to find topological solutions, whose classical limit depends on all values of the angular momentum. We appeal to another method, based on projective modules, to find noncommutative monopoles inside the matrix model, tending to classical monopoles of the sphere. The projective modules are built making use of the the principal Hopf fibration $\pi : S^3 \to S^2$, and its extension to the fuzzy sphere case. The link with the matrix models is possible only if the projectors can be recast in the form of a ket-bra combination. Once that the resulting vector valued operator is known, it is possible to construct topological solutions of the matrix model at the noncommutative level.

1 Introduction

The properties of noncommutative gauge theory [1]-[3] defined on a noncommutative manifold have recently received growing attention in the physical literature. The simplest noncommutative manifold to work with is the fuzzy sphere, which has the important simplification of admitting only a finite number of states [4]-[10]. As a consequence it is possible to reach many interesting results that can be taken as a paradigm for more general cases. One of the best definitions of noncommutative gauge theory on a fuzzy sphere turns out to be introducing a matrix model [11]-[21] which contains the fuzzy sphere algebra as a classical solution [22]-[25]. Then expanding the matrix model around a classical solution, the fluctuations on the background can be interpreted as fields of noncommutative gauge theories.

The matrix model contains many more classical solutions than the fuzzy sphere, so our first investigation was to look for new classes of solutions and answer the question whether these new classes of solutions lead to solitons for the noncommutative gauge theory on the fuzzy sphere.

The matrix model we considered depends on two coupling constants g^2 and λ. Starting with an ansatz

$$X_i = A_i^j L_j \tag{1.1}$$

we found in [26] a class of new solutions besides the fuzzy sphere for $\lambda < 1$. In general the difference between the fuzzy sphere background $\hat{x}_i$ and the new class of classical solutions X_i (1.1) is such that it cannot be interpreted as fluctuation of a noncommutative gauge theory, which must go to zero as $\frac{1}{N}$, in the $N \to \infty$ limit, thus preventing the use of them to construct a soliton.

Fortunately we found fruitful in [27] exploring the neighborhood of $\lambda = \frac{1}{2}$, (precisely $\lambda \sim \frac{1}{2} + \frac{\lambda_0}{N}$) to find a good fluctuation, i.e. a field of noncommutative gauge theory tending to a classical soliton of the sphere in the $N \to \infty$ limit.

The gauge field and the scalar field derived from the fluctuation, defined as the difference of the two solutions of the matrix model, are true noncommutative solitons of the fuzzy sphere and satisfy the classical equations of motion in the $N \to \infty$ limit. In particular the equation of motion for the scalar field is equal to the eigenvalue problem for the L^2 operator, whose possible solutions are the spherical harmonics $Y_l^m(\Omega)$. In our case, the ansatz (1.1) is related to the choice $l = 2$. Noncommutative solitons with higher spherical harmonics $l > 2$ are possible for models with λ around the classical value $\lambda = \frac{3-l}{2}$ or $\lambda = \frac{4+l}{2}$. Correspondingly one should start from solutions of matrix model obtained with more general ansatze

$$X_i = A_i^{j_1,\ldots,j_{l-1}} L_{j_1} \ldots L_{j_{l-1}} \tag{1.2}$$

We show that also the case $l = 0$ follows the same pattern of the $\lambda = \frac{1}{2}$ case and produces noncommutative solitons around $\lambda = 2$. This value of λ is rather special, since in the classical limit one expects to find two distinct classes of solutions, one with a constant scalar field and the other one corresponding to the monopole solution.

Therefore one expects to find generalizations of the monopoles, which should be interpreted as topologically nontrivial configurations over a fuzzy sphere. The method outlined before is not indicated to find such configurations since the ansatz we choose truncates the angular momentum to a fixed limiting value l in the $N \to \infty$ limit, while the monopole class requires contribution from all values of the angular momentum.

To find noncommutative generalizations of the monopoles, we have followed another strategy, based on projective modules [28]-[31]. The Serre-Swan theorem permits the study of nontrivial topologies for vector bundles in terms of projective modules, at a classical level [32]-[34]. We have attempted to make an analogous bridge at a noncommutative level, to reach the noncommutative monopoles on a fuzzy sphere [35]. In ref. [31] we have identified the nontrivial projective modules on a fuzzy sphere, by using the Hopf principal fibration $\pi : S^3 \to S^2$ on the two-sphere, with $U(1)$ as a structure group. Then in ref. [35] we have deconstructed the projectors in terms of more fundamental vector-valued fields and defined the corresponding topologically nontrivial matrix model configurations, identifying the class of models for which the projectors of Ref. [31] are indeed solution of the $Y - M$ equations of motion.

2 Properties of the Fuzzy Sphere

In this review we will treat mainly noncommutative solutions of Yang-Mills theory on the fuzzy sphere. In ref. [4] it was shown how to truncate consistently the geometry of the sphere S^2 to a matrix geometry, represented by the M_{N+1} algebra of $(N+1) \otimes (N+1)$ complex matrices.

The classical geometry of the sphere is defined by the algebra $\mathcal{C}(S^2)$ of smooth functions on S^2. An approximation of this function space can be obtained by limiting the polynomial expansion of the smooth functions on S^2 to a finite N number of terms, i.e. neglecting the higher order terms of the expansion.

In order to turn this finite approximation to a true algebra we must substitute the commuting coordinates of the sphere with a noncommutative manifold represented by the following algebra

$$[\hat{x}^i, \hat{x}^j] = i\rho\epsilon^{ijk}\hat{x}^k \quad i,j,k = 1,2,3 \tag{2.1}$$

In general $\hat{x}^i$ are realized by $(N+1) \times (N+1)$ hermitian matrices, identified with the generators of the $(N+1)$-dimensional irreducible representation of $SU(2)$

$$\hat{x}^i = \rho\hat{L}^i \tag{2.2}$$

The basic constraint of the sphere, the radius of the sphere, is proportional to the Casimir of the representation of $SU(2)$

$$\hat{x}^i\hat{x}^i = R^2 = \rho^2\hat{L}^i\hat{L}^i = \rho^2\frac{N(N+2)}{4} \tag{2.3}$$

By requiring that the radius of the sphere is fixed in the large N limit, we obtain a commutative limit for the fuzzy sphere algebra

$$R = \text{fixed} \quad \rho \to 0 \quad (N \to \infty) \tag{2.4}$$

where $\hat{x}^i$ become the normal coordinates on the sphere x^i:

$$\begin{aligned} x^1 &= Rsin\theta cos\phi \\ x^2 &= Rsin\theta sin\phi \\ x^3 &= Rcos\theta \end{aligned} \tag{2.5}$$

This embedding produces the usual metric tensor of the sphere:

$$ds^2 = R^2(d\theta^2 + \sin^2\theta d\phi^2) = R^2 g_{ab} d\sigma_a d\sigma_b \tag{2.6}$$

Another useful representation of the fuzzy sphere algebra is obtained by observing that the classical x_i coordinates are related to a couple of complex coordinates, normalized to live on S^3, by the Hopf principal fibration.

The $SU(2)$ Lie algebra is then realized by promoting the two complex coordinates to a couple of independent oscillators:

$$\begin{aligned} z_i &\rightarrow a_i \qquad [a_i, a_j^\dagger] = \delta_{ij} \\ \hat{x}_1 &= \frac{\hat{\rho}}{2}(a_0 a_1^\dagger + a_1 a_0^\dagger) \\ \hat{x}_2 &= i\frac{\hat{\rho}}{2}(a_0 a_1^\dagger - a_1 a_0^\dagger) \\ \hat{x}_3 &= \frac{\hat{\rho}}{2}(a_0 a_0^\dagger - a_1 a_1^\dagger) \\ \hat{N} &= a_0^\dagger a_0 + a_1^\dagger a_1. \end{aligned} \tag{2.7}$$

There is another operator, which can be defined through the oscillators, the total oscillator number $\hat{N} = N$, whose eigenvalue on a definite $SU(2)$ representation coincides with the cutoff parameter of the function space on the fuzzy sphere algebra. The operator $\hat{\rho}$ is equal to ρ for irreducible $SU(2)$ representations with rank $(N+1)$.

It is clear from this construction that the fuzzy sphere function space is finite-dimensional, in contrast to what happens in the commutative limit. The number of independent functions on the fuzzy sphere is $(N+1)^2$, which is exactly the number of parameters of a $(N+1) \times (N+1)$ hermitian matrix.

Since the polynomial functions of the coordinates x^i can be related to the classical spherical harmonics, an alternative way of thinking is introducing a cutoff parameter N for the angular momentum of the spherical harmonics. An ordinary function belonging to $\mathcal{C}(S^2)$ on the sphere can be expanded in terms of the spherical harmonics:

$$a(\Omega) = \sum_{l=0}^{\infty} \sum_{m=-l}^{l} a_{lm} Y_{lm}(\Omega) \tag{2.8}$$

The $Y_{lm}(\Omega)$ form a basis of the classical infinite-dimensional function space. By introducing a cutoff N on the number l, the number of independent functions reduces to $\sum_{l=0}^{N}(2l+1) = (N+1)^2$, remembering that at the same time we must substitute the elements of the classical basis with elements of the fuzzy sphere algebra. The noncommutative analogue of the spherical harmonics can be determined by looking at their classical form

$$Y_{lm} = R^{-l} \sum_a f^{(lm)}_{a_1,a_2,\dots,a_l} x^{a_1} \dots x^{a_l} \tag{2.9}$$

and $f_{a_1,a_2,\dots,a_l}$ is a traceless and symmetric tensor. The normalization of the spherical harmonics is fixed by

$$\int \frac{d\Omega}{4\pi} Y^*_{l'm'} Y_{lm} = \delta_{ll'} \delta_{mm'} \tag{2.10}$$

The corresponding noncommutative spherical harmonics $\hat{Y}_{lm}$ are $(N+1) \times (N+1)$ hermitian matrices ,

$$\hat{Y}_{lm} = R^{-l} \sum_a f^{(lm)}_{a_1,a_2,..,a_l} \hat{x}_{a_1}....\hat{x}_{a_l} \tag{2.11}$$

defined by the same symmetric tensor $f^{(lm)}_{a_1,a_2,..,a_l}$. A Weyl type ordering is implicit into this definition, due to the symmetry of the indices [5].

Also the normalization of the noncommutative spherical harmonics has to be changed with respect to the classical formula (2.10)

$$\frac{1}{N+1} Tr(\hat{Y}^\dagger_{l'm'} \hat{Y}_{lm}) = \delta_{l'l} \delta_{m'm} \tag{2.12}$$

An alternative definition of this noncommutative function space is using the classical basis of spherical harmonics endowed with a noncommutative star product analogous to the Moyal star product for the plane.

A matrix on the fuzzy sphere

$$\hat{a} = \sum_{l=0}^{N} \sum_{m=-l}^{l} a_{lm} \hat{Y}_{lm} \qquad a^*_{lm} = a_{l-m} \tag{2.13}$$

can be substituted by a representative (an ordinary function on the commutative sphere), with a cutoff on the angular momentum:

$$a(\Omega) = \sum_{l=0}^{N} \sum_{m=-l}^{l} a_{lm} Y_{lm}(\Omega) \tag{2.14}$$

where

$$a(\Omega) = \frac{1}{N+1} \sum_{l=0}^{N} \sum_{m=-l}^{l} Tr(\hat{Y}^\dagger_{lm} \hat{a}) Y_{lm}(\Omega) \tag{2.15}$$

The closure of the function algebra is obtained by defining a star product on the commutative sphere which makes use of the ordinary product of matrices:

$$\begin{aligned} \hat{a}\hat{b} \quad &\longrightarrow \quad a * b \\ a(\Omega) * b(\Omega) \quad &= \quad \frac{1}{N+1} \sum_{l=0}^{N} \sum_{m=-l}^{l} Tr(\hat{Y}^\dagger_{lm} \hat{a}\hat{b}) Y_{lm}(\Omega) \end{aligned} \tag{2.16}$$

Then we stress again that the cutoff in the classical function space is consistent only with the star product of classical functions, not with the commutative one.

Derivative operators can be constructed by the adjoint action of $\hat{L}_i$:

$$Ad(\hat{L}_i)\hat{a} = \sum_{l=0}^{N}\sum_{m=-l}^{l} a_{lm}\,[\hat{L}_i, \hat{Y}_{lm}] \tag{2.17}$$

In the classical limit $Ad(\hat{L}_i)$ tends to the Lie derivative on the sphere:

$$Ad(\hat{L}_i) \to L_i = \frac{1}{i}\epsilon_{ijk}x_j\partial_k \tag{2.18}$$

where we can expand the classical Lie derivative L_i in terms of the Killing vectors of the sphere

$$L_i = -ik_i^a\partial_a \tag{2.19}$$

In terms of k_i^a we can form the metric tensor $g_{ab} = k_a^i k_b^i$.
In particular the analogue of the Laplacian on the fuzzy sphere is given by :

$$\begin{aligned}\frac{1}{R^2}Ad(\hat{L}^2)\hat{a} &= \frac{1}{R^2}\sum_{l=0}^{N}\sum_{m=-l}^{l} a_{lm}[\hat{L}_i, [\hat{L}_i, \hat{Y}_{lm}]] = \\ &= \sum_{l=0}^{N}\sum_{m=-l}^{l} \frac{l(l+1)}{R^2}a_{lm}\hat{Y}_{lm}\end{aligned} \tag{2.20}$$

Trace over matrices can be mapped to the integration over functions:

$$\frac{1}{N+1}Tr(\hat{a}) \to \int\frac{d\Omega}{4\pi}a(\Omega) \tag{2.21}$$

3 Gauge Theory on the Fuzzy Sphere from Matrix Model

In order to define the noncommutative gauge theory on a fuzzy sphere we follow mainly [22]-[23], using the following matrix model depending on two coupling constants, g^2 and λ:

$$S(\lambda) = S_0 + \lambda S_1 = -\frac{1}{g^2}Tr[\frac{1}{4}[X_i, X_j][X_i, X_j] - \frac{2}{3}i\lambda\rho\epsilon^{ijk}X_iX_jX_k + \rho^2(1-\lambda)X_iX_i]. \tag{3.1}$$

The action $S(\lambda)$ is the most general action which contains the fuzzy sphere algebra as a classical solution. To recover gauge theory on a fuzzy sphere we must develop the matrix X_i as a sum of the classical background $\hat{x}_i$ and the fluctuation around the background, which represents the connection of the noncommutative gauge theory:

$$X_i = \hat{x}_i + \rho R\hat{a}_i \tag{3.2}$$

As a consequence a $U(1)$ noncommutative gauge theory on the fuzzy sphere is naturally introduced through the fluctuation $\hat{a}_i$. More generally a $U(m)$ gauge group is simply obtained by extending the matrix model in order to permit the $U(m)$ background classical solution:

$$\hat{x}_i \;\rightarrow\; \hat{x}_i \otimes 1_m \tag{3.3}$$

In this case the fluctuations $\hat{a}_i$ are generalized as follows:

$$\hat{a}_i \rightarrow \sum_{a=1}^{m^2} \hat{a}_i^a \otimes T^a \tag{3.4}$$

where $T^a(a = 1, 2, ..., m^2)$ denote the generators of $U(m)$.

The matrix model action (3.1) contains the ordinary gauge transformations, being invariant under the following unitary transformation

$$X_i \rightarrow U^{-1} X_i U \tag{3.5}$$

which is the usual transformation for a covariant derivative (in the classical limit, the matrix model variable X_i has the meaning of a covariant derivative).

In fact by developing U for an infinitesimal transformation:

$$U \sim 1 + i\hat{\lambda} \quad \hat{\lambda} = \sum_{l=0}^{N} \sum_{m=-l}^{l} \lambda_{lm} \hat{Y}_{lm} \tag{3.6}$$

the fluctuations around the fixed background transforms as

$$\hat{a}_i \;\rightarrow\; \hat{a}_i - \frac{i}{R}[\hat{L}_i, \hat{\lambda}] + i[\hat{\lambda}, \hat{a}_i] \tag{3.7}$$

which can be mapped to a standard local gauge symmetry with the star-product substituting the ordinary product:

$$a_i(\Omega) \;\rightarrow\; a_i(\Omega) - \frac{i}{R} L_i \lambda(\Omega) + i[\lambda(\Omega), a_i(\Omega)]_* \tag{3.8}$$

The symbol $()_*$ means that the product is to be considered as a star product.

The corresponding field strength on the sphere is given by

$$\begin{aligned} \hat{F}_{ij} &= \frac{1}{\rho^2 R^2}([X_i, X_j] - i\rho\epsilon_{ijk} X_k) \\ &= [\frac{\hat{L}_i}{R}, \hat{a}_j] - [\frac{\hat{L}_j}{R}, \hat{a}_i] + [\hat{a}_i, \hat{a}_j] - \frac{i}{R}\epsilon_{ijk}\hat{a}_k \end{aligned} \tag{3.9}$$

Its expression in terms of classical functions is

$$F_{ij}(\Omega) = \frac{1}{R} L_i a_j(\Omega) - \frac{1}{R} L_j a_i(\Omega) + [a_i(\Omega), a_j(\Omega)]_* - \frac{i}{R}\epsilon_{ijk} a_k \tag{3.10}$$

F_{ij} is gauge covariant even in the $U(1)$ case, as it is manifest from the viewpoint of the matrix model.

It is possible to define a scalar field from the matrix variable X_i, belonging to the adjoint representation (a sort of gauge field) defined as:

$$\hat{\phi} = \frac{1}{2\rho R}(X_i X_i - \hat{x}_i \hat{x}_i) = \frac{1}{2}(\hat{x}_i \hat{a}_i + \hat{a}_i \hat{x}_i + \rho R \hat{a}_i \hat{a}_i) \tag{3.11}$$

However at the noncommutative level it is meaningless to separate the contribution of the scalar field from that of the gauge field. Only in the classical limit the action can be represented as a sum of both contributions.

To study the commutative limit of the model (3.1), we must first map the matrix model action to a classical action with the star product and then perform the following large N R fixed limit:

$$R = \text{fixed}, \quad g_{YM}^2 = \frac{4\pi g^2}{(N+1)\rho^4 R^2} = \text{fixed} \quad N \to \infty \tag{3.12}$$

In the commutative limit, the star product becomes the commutative product making separable the scalar field ϕ from the gauge field as in

$$R a_i(\Omega) = k_i^a b_a(\Omega) + \frac{x_i}{R}\phi(\Omega) \tag{3.13}$$

where b_a is a true gauge field on the sphere. The field strength F_{ij} can be expanded in terms of the gauge field b_a and the scalar field ϕ as follows :

$$F_{ij}(\Omega) = \frac{1}{R^2} k_i^a k_j^b F_{ab} + \frac{i}{R^3}\epsilon_{ijk} x_k \phi + \frac{1}{R^3} x_j k_i^a D_a \phi - \frac{1}{R^3} x_i k_j^a D_a \phi \tag{3.14}$$

where $F_{ab} = -i(\partial_a b_b - \partial_b b_a) + [b_a, b_b]$ and $D_a = -i\partial_a + [b_a, .]$.

Let us first work out the classical limit for the S_0 part of the action, defined implicitly in (3.1) as

$$S_0 = -\frac{1}{g^2} Tr(\frac{1}{4}[X_i, X_j][X_i, X_j] + \rho^2 X_i X_i) \tag{3.15}$$

Firstly the action S_0 is mapped, following again reference [22]-[23], through the map (2.16) to the following field theory action with the star product as follows

$$\begin{aligned} S_0 = & - \frac{R^2}{4 g_{YM}^2} Tr \int d\Omega (F_{ij} F_{ij}) \\ & - \frac{3i}{2 g_{YM}^2} \epsilon_{ijk} Tr \int d\Omega ((L_i a_j) a_k + \frac{R}{3}[a_i, a_j] a_k - \frac{i}{2}\epsilon_{ijl} a_l a_k)_* \\ & - \frac{\pi}{g_{YM}^2} \frac{N(N+2)}{2R^2} \end{aligned} \tag{3.16}$$

The classical limit, realized as in (3.12), produces the following contribution

$$\begin{aligned} S_0 &= -\frac{1}{4g_{YM}^2R^2}Tr\int d\Omega(k_i^ak_j^bk_i^ck_j^dF_{ab}F_{cd} + 2ik_i^ak_j^bF_{ab}\epsilon_{ijk}\frac{x_k}{R}\phi \\ &+ 2k_i^ak_i^b(D_a\phi)(D_b\phi) - 2\phi^2) \\ &- \frac{3}{2g_{YM}^2R^2}Tr\int d\Omega(i\epsilon_{ijk}k_i^ak_j^bF_{ab}\frac{x_k}{R}\phi - \phi^2) \\ &= -\frac{1}{4g_{YM}^2R^2}Tr\int d\Omega(F_{ab}F^{ab} + 8i\frac{\epsilon^{ab}}{\sqrt{g}}F_{ab}\phi + 2(D_a\phi)(D^a\phi) - 8\phi^2) \end{aligned} \tag{3.17}$$

Analogously the action S_1

$$S_1 = \frac{\rho}{g^2}Tr[\frac{2}{3}i\epsilon^{ijk}X_iX_jX_k + \rho X^iX_i]. \tag{3.18}$$

is mapped to the following field theory action as follows:

$$\begin{aligned} S_1 &= \frac{i}{g_{YM}^2}\epsilon_{ijk}Tr\int d\Omega((L_ia_j)a_k + \frac{R}{3}[a_i,a_j]a_k - \frac{i}{2}\epsilon_{ijl}a_la_k)_* \\ &+ \frac{\pi}{3g_{YM}^2}\frac{N(N+2)}{R^2} \end{aligned} \tag{3.19}$$

The commutative limit simplifies the action (3.19) to

$$\begin{aligned} S_1 &= \frac{1}{g_{YM}^2R^2}Tr\int d\Omega(i\epsilon_{ijk}k_i^ak_j^bF_{ab}\frac{x^k}{R}\phi - \phi^2) = \\ &= \frac{1}{g_{YM}^2R^2}Tr\int d\Omega\left(\frac{i\epsilon^{ab}}{\sqrt{g}}F_{ab}\phi - \phi^2\right) \end{aligned} \tag{3.20}$$

By combining the results (3.17) and (3.20) the complete action $S(\lambda)$ becomes

$$\begin{aligned} S(\lambda) &= S_0 + \lambda S_1 = \\ &- \frac{1}{4g_{ym}^2R^2}\int d\Omega[F_{ab}F^{ab} - 2\partial_a\phi\partial_a\phi + (8-4\lambda)(\frac{i\epsilon^{ab}}{\sqrt{g}}F_{ab}\phi - \phi^2)]. \end{aligned} \tag{3.21}$$

As in the Bogomolnyi trick, we can isolate an equation of motion directly at the level of action as a quadratic term

$$\begin{aligned} S(\lambda) &= -\frac{1}{4g_{YM}^2}\int d\Omega[(F_{ab} + (4-2\lambda)i\epsilon_{ab}\phi\sqrt{g})(F^{ab} + (4-2\lambda)i\epsilon^{ab}\frac{\phi}{\sqrt{g}}) + \\ &- 2\partial_a\phi\partial^a\phi + (8(2-\lambda)^2 - 4(2-\lambda))\phi^2]. \end{aligned} \tag{3.22}$$

The auxiliary scalar field ϕ can be decoupled from pure Yang-Mills theory for the particular value $\lambda = 2$, in which case the action reduces to :

$$S(2) = -\frac{1}{4g_{YM}^2}\int d\Omega(F_{ab}F^{ab} - 2\partial_a\phi\partial^a\phi)$$
$$d\Omega = \sqrt{g}d\theta d\phi = sin\theta d\theta d\phi \qquad F^{ab} = g^{aa'}g^{bb'}F_{a'b'}. \tag{3.23}$$

i.e. there is no mixing term between ϕ and F_{ab}.

4 Soliton Solutions

The equations of motion for the matrix model action $S(\lambda)$ contain other solutions rather than the fuzzy sphere, as found in [26]. These solutions can be used to construct solitons solutions for the fuzzy sphere as done in [27].

Let us recall the whole construction, starting from the general equations of motion of the matrix model (3.1)

$$[X_j,[X_i,X_j]] - i\rho\lambda\epsilon^{ijk}[X_j,X_k] + 2\rho^2(1-\lambda)X_i = 0. \tag{4.1}$$

A simplified ansatz is needed to reduce these general equations in a solvable set. A consistent choice turns out to be the following

$$X_i = A_i^k L_k \tag{4.2}$$

where the coefficients A_i^k represent three arbitrary vectors A_i^1, A_i^2, A_i^3 that are constrained by the equations

$$\begin{aligned}
&(2\rho^2(1-\lambda) - \beta^2 - \gamma^2)A_i^1 + (A^1\cdot A^2)A_i^2 + (A^1\cdot A^3)A_i^3 + 2\lambda\rho\epsilon_{ijk}A_j^2A_k^3 = 0\\
&(2\rho^2(1-\lambda) - \alpha^2 - \gamma^2)A_i^2 + (A^1\cdot A^2)A_i^1 + (A^2\cdot A^3)A_i^3 + 2\lambda\rho\epsilon_{ijk}A_j^3A_k^1 = 0\\
&(2\rho^2(1-\lambda) - \alpha^2 - \beta^2)A_i^3 + (A^1\cdot A^3)A_i^1 + (A^2\cdot A^3)A_i^2 + 2\lambda\rho\epsilon_{ijk}A_j^1A_k^2 = 0
\end{aligned} \tag{4.3}$$

having used the following parameterizations

$$(A_i^1)^2 = \alpha^2 \quad (A_i^2)^2 = \beta^2 \quad (A_i^3)^2 = \gamma^2. \tag{4.4}$$

The three vectors can be oriented in the 3d internal space as follows:

$$\begin{aligned}
&A_x^1 = \alpha \quad A_y^1 = A_z^1 = 0\\
&A_x^2 = \beta cos\theta_{12} \quad A_y^2 = \beta sin\theta_{12} \quad A_z^2 = 0\\
&A_x^3 = \gamma cs\theta_{13} \quad A_y^3 = \gamma sin\theta_{13}sin\phi \quad A_z^3 = \gamma sin\theta_{13}cos\phi.
\end{aligned} \tag{4.5}$$

Apart from the fuzzy sphere, there is another class of solutions to the equations (4.1), characterized by the unique constraint:

$$\alpha^2 + \beta^2 + \gamma^2 = 4\rho^2(1-\lambda) + 4\lambda^2\rho^2 \tag{4.6}$$

together with the following inequalities

$$\alpha^2 \geq 2\rho^2(1-\lambda) \quad \beta^2 \geq 2\rho^2(1-\lambda) \quad \gamma^2 \geq 2\rho^2(1-\lambda). \tag{4.7}$$

Then the vector X_i is completely determined in terms of (4.6) and (4.7)

$$\begin{aligned}
cos\phi &= -\frac{2\lambda\alpha\rho}{\sqrt{(2\rho^2(1-\lambda)-\beta^2)(2\rho^2(1-\lambda)-\gamma^2)+4\lambda^2\alpha^2\rho^2}} \\
sin\phi &= \sqrt{\frac{(2\rho^2(1-\lambda)-\beta^2)(2\rho^2(1-\lambda)-\gamma^2)}{(2\rho^2(1-\lambda)-\beta^2)(2\rho^2(1-\lambda)-\gamma^2)+4\lambda^2\alpha^2\rho^2}} \\
cos\theta_{12} &= -\frac{1}{\alpha\beta}\sqrt{(2\rho^2(1-\lambda)-\alpha^2)(2\rho^2(1-\lambda)-\beta^2)} \\
sin\theta_{12} &= \frac{1}{\alpha\beta}\sqrt{2\rho^2(1-\lambda)(\alpha^2+\beta^2-2\rho^2(1-\lambda))} \\
cos\theta_{13} &= -\frac{1}{\alpha\gamma}\sqrt{(2\rho^2(1-\lambda)-\alpha^2)(2\rho^2(1-\lambda)-\gamma^2)} \\
sin\theta_{13} &= -\frac{1}{\alpha\gamma}\sqrt{2\rho^2(1-\lambda)(\alpha^2+\gamma^2-2\rho^2(1-\lambda)} \\
cos\theta_{23} &= -\frac{1}{\beta\gamma}\sqrt{2\rho^2(1-\lambda)-\beta^2)(2\rho^2(1-\lambda)-\gamma^2)} \\
sin\theta_{23} &= -\frac{1}{\beta\gamma}\sqrt{2\rho^2(1-\lambda)(\beta^2+\gamma^2-2\rho^2(1-\lambda))}
\end{aligned} \tag{4.8}$$

where $cos\theta_{23} = cos\theta_{12}cos\theta_{13} + sin\theta_{12}sin\theta_{13}sin\phi$.

Therefore it is possible to generate many solutions of the matrix model (3.1), but there is a restriction to apply calling a noncommutative soliton a configuration which tends to a classical soliton of the sphere in the $N \to \infty$ limit.

The soliton configuration comes from the so called fluctuation, which is defined in (3.2) as the difference between the fuzzy sphere $\hat{x}^i$ background and the new class of classical solutions X_i. It is not difficult to realize that in order to recover a well-defined classical limit, the fluctuations must go to zero as $\frac{1}{N}$, in the $N \to \infty$ limit. Instead in the general class of solutions (4.6), (4.7), (4.8) X_i differs significantly from the fuzzy sphere $\hat{x}_i$ and cannot be used to define a consistent connection as in (3.2).

However since the difference of the classical action computed on the class of solutions (4.6) and the classical action computed on the fuzzy sphere goes like $(\lambda - \frac{1}{2})^3$, we can make it negligible in the $N \to \infty$ limit if λ scales as

$$\lambda = \frac{1}{2} + \rho\lambda_0. \tag{4.9}$$

Implementing this further constraint in the general class of solutions we are able to produce non trivial connections $\hat{a}_i$, which then describe noncommutative solitons for the fuzzy sphere.

Firstly we must make these two constraints compatible

$$\alpha^2+\beta^2+\gamma^2 = 4\rho^2(1-\lambda)+4\rho^2\lambda^2$$
$$\lambda = \frac{1}{2}+\rho\lambda_0 \tag{4.10}$$

A possible choice is obtained by parameterizing α, β, γ as:

$$\alpha = \rho + \alpha_0\rho^2$$
$$\beta = \rho + \beta_0\rho^2$$
$$\gamma = \rho + \gamma_0\rho^2 \tag{4.11}$$

giving rise to the following conditions

$$\alpha_0+\beta_0+\gamma_0 = 0$$
$$\alpha_0^2+\beta_0^2+\gamma_0^2 = 4\lambda_0^2. \tag{4.12}$$

A general solution of these conditions turns out to be

$$\alpha_0 = 2\sqrt{\frac{2}{3}}\lambda_0 sin(\frac{\pi}{3}-\theta)$$
$$\beta_0 = 2\sqrt{\frac{2}{3}}\lambda_0 sin\theta$$
$$\gamma_0 = -2\sqrt{\frac{2}{3}}\lambda_0 sin(\frac{\pi}{3}+\theta). \tag{4.13}$$

More work is needed to satisfy the inequalities (4.7) which are necessary to make consistent the solution (4.8). By defining

$$sin\alpha_0 = \frac{1}{2}\sqrt{\frac{3}{2}} \tag{4.14}$$

these inequalities imply that

$$sin(\frac{\pi}{3}-\theta) \geq -sin\alpha_0 \quad sin\theta \geq -sin\alpha_0 \quad sin(\frac{\pi}{3}+\theta) \leq sin\alpha_0. \tag{4.15}$$

Fortunately there is a narrow window of values for θ which satisfies all these inequalities, if $\lambda_0 > 0$

$$\frac{4}{3}\pi - \alpha_0 < \theta < \pi + \alpha_0. \tag{4.16}$$

It is possible to fix the value of θ to a particular value to simplify the notation in the following

$$\theta = \frac{7}{6}\pi \tag{4.17}$$

Summarizing we have found a solution to the constraints (4.6) and (4.7)

$$\begin{aligned}\alpha &= \rho(1 - \sqrt{\frac{2}{3}}\lambda_0\rho) \\ \beta &= \rho(1 - \sqrt{\frac{2}{3}}\lambda_0\rho) \\ \gamma &= \rho(1 + 2\sqrt{\frac{2}{3}}\lambda_0\rho).\end{aligned} \tag{4.18}$$

which defines a particular noncommutative soliton tending to a classical soliton around the value $\lambda = \frac{1}{2}$ since the fluctuations respect the dependence on ρ of equation (3.2).

To make clear the physical meaning of this solution we now look for the classical limit of it and verify that it is a nontrivial solution of the classical limit of the action $S(\lambda)$, i.e. a classical soliton for the sphere. To this end we recall the classical limit of the action $S(\lambda)$, as determined in equation (3.22), from which the resulting equations of motion are

$$\begin{aligned}&F_{ab} + (4 - 2\lambda)i\epsilon_{ab}\phi\sqrt{g} = 0 \\ &\frac{1}{\sqrt{g}}\partial_a(\sqrt{g}g^{ab}\partial_b)\phi = L_iL_i\phi = [(4-2\lambda)^2 - 2(2-\lambda)]\phi.\end{aligned} \tag{4.19}$$

We can recognize in the second equation the eigenvalue problem for the L_iL_i operator which admits nontrivial solutions if and only if

$$L_iL_i\phi = l(l+1)\phi \tag{4.20}$$

the coefficient in front of ϕ in the second member is equal to $l(l+1)$. By imposing that

$$l(l+1) = (4-2\lambda)^2 - 2(2-\lambda) \tag{4.21}$$

the only classical models which admit nontrivial solutions are for

$$\lambda = \frac{3-l}{2} \quad \text{or} \quad \lambda = \frac{l+4}{2}. \tag{4.22}$$

It is easy to recognize that our ansatz (4.2) belongs to the case $l = 2$ which corresponds to the classical value $\lambda = \frac{1}{2}$, in complete agreement with the development (4.10).

To check that the classical limit of the fluctuations a_i satisfies the classical equations of motion for $\lambda = \frac{1}{2}$, let us extract a_i in terms of the solution for X_i (4.8), developed in power of ρ, keeping also the first subleading contribution:

$$A_x^1 = \rho(1 - \sqrt{\frac{2}{3}}\lambda_0\rho)$$

$$
\begin{aligned}
A_x^2 &= -2\lambda_0(1-\sqrt{\frac{2}{3}})\rho^2(1-\sqrt{\frac{2}{3}}\lambda_0\rho)+O(\rho^4) \\
A_y^2 &= \rho(1-\sqrt{\frac{2}{3}}\lambda_0\rho-2\lambda_0^2(1-\sqrt{\frac{2}{3}})^2\rho^2)+O(\rho^4) \\
A_x^3 &= -2\lambda_0\sqrt{(1-\sqrt{\frac{2}{3}})(1+2\sqrt{\frac{2}{3}})}\rho^2(1+2\sqrt{\frac{2}{3}}\lambda_0\rho)+O(\rho^4) \\
A_y^3 &= 2\lambda_0\sqrt{(1-\sqrt{\frac{2}{3}})(1+2\sqrt{\frac{2}{3}})}\rho^2(1+2\sqrt{\frac{2}{3}}\lambda_0\rho)+O(\rho^4) \\
A_z^3 &= \rho(1+2\sqrt{\frac{2}{3}}\lambda_0-4\lambda_0^2(1-\sqrt{\frac{2}{3}})(1+2\sqrt{\frac{2}{3}})\rho^2)+O(\rho^4).
\end{aligned} \tag{4.23}
$$

Strictly speaking the classical fluctuations a_i are deducible from (4.23) as follows:

$$
\begin{aligned}
Ra^1 &= -\sqrt{\frac{2}{3}}\lambda_0 x^1 \\
Ra^2 &= -2\lambda_0(1-\sqrt{\frac{2}{3}})x_1-\sqrt{\frac{2}{3}}\lambda_0 x_2 \\
Ra_3 &= -2\lambda_0\sqrt{(1-\sqrt{\frac{2}{3}})(1+2\sqrt{\frac{2}{3}})}(x_1-x_2)+2\sqrt{\frac{2}{3}}\lambda_0 x_3.
\end{aligned} \tag{4.24}
$$

The classical scalar field, obtained from this particular fluctuation a_i , is :

$$
R\phi(\Omega) = x_i \cdot Ra_i \tag{4.25}
$$

therefore

$$
\begin{aligned}
R\phi(\Omega) &= \sqrt{\frac{2}{3}}\lambda_0(x^1)^2-2\lambda_0(1-\sqrt{\frac{2}{3}})x^1x^2-\sqrt{\frac{2}{3}}(x^2)^2 \\
&- 2\lambda_0\sqrt{(1-\sqrt{\frac{2}{3}})(1+2\sqrt{\frac{2}{3}})}(x^1-x^2)x^3 \\
&+ 2\sqrt{\frac{2}{3}}\lambda_0(x^3)^2
\end{aligned} \tag{4.26}
$$

or, in usual spherical coordinates,

$$
\begin{aligned}
\phi(\Omega) &= R[\lambda_0\sqrt{\frac{2}{3}}(2cos^2\theta-sin^2\theta)-2\lambda_0(1-\sqrt{\frac{2}{3}})sin^2\theta sin\phi cos\phi \\
&+ 2\lambda_0\sqrt{(1-\sqrt{\frac{2}{3}})(1+2\sqrt{\frac{2}{3}})}sin\theta cos\theta(sin\phi-cos\phi)].
\end{aligned} \tag{4.27}
$$

Using the explicit form of the Killing vector k_i^a, we can deduce the two components of the $U(1)$ gauge field as:

$$b_\theta = -Ra^1 sin\phi + Ra^2 cos\phi = \\ = -2R\lambda_0(1-\sqrt{\frac{2}{3}})sin\theta cos^2\phi \quad (4.28)$$

and

$$b_\phi = R[2\lambda_0\sqrt{(1-\sqrt{\frac{2}{3}})(1+2\sqrt{\frac{2}{3}})}sin^3\theta(sin\phi - cos\phi) \\ + 3\lambda_0\sqrt{\frac{2}{3}}cos\theta sin^2\theta + 2\lambda_0(1-\sqrt{\frac{2}{3}})sin^2\theta cos\theta sin\phi cos\phi]. \quad (4.29)$$

Now we are ready to verify that the soliton solutions (4.27), (4.28) and (4.29) are non trivial solutions of the classical equations of motion (4.19) on the sphere. Firstly we notice from the formula (4.27) that the scalar field $\phi(\Omega)$ is a combination of Y_2^m spherical harmonics, and this is enough to solve the second equation of motion (4.19) for $\lambda = \frac{1}{2}$.

To check also the other equation we need to compute $F_{\theta\phi}$

$$F_{\theta\phi} = -i(\partial_\theta b_\phi - \partial_\phi b_\theta) = \\ - 3isin\theta R[2\lambda_0\sqrt{(1-\sqrt{\frac{2}{3}})(1+2\sqrt{\frac{2}{3}})}sin\theta cos\theta(sin\phi - cos\phi) \\ + \lambda_0\sqrt{\frac{2}{3}}sin\theta(2cos^2\theta - sin^2\theta - 2\lambda_0(1-\sqrt{\frac{2}{3}})sin^2\theta cos\phi sin\phi] = \\ = -3isin\theta\phi(\Omega)\epsilon_{\theta\phi} \quad (4.30)$$

which satisfies the other equations of motion for $\lambda = \frac{1}{2}$. Our solution is a noncommutative soliton for the fuzzy sphere which, in the $N \to \infty$ limit, corresponds to a classical soliton (with spherical harmonics $l = 2$) for the sphere with the classical action $S(\frac{1}{2})$.

From this construction it should be clear that to find noncommutative solitons with higher spherical harmonics $l > 2$ one should explore the neighborhood of the models

$$\lambda = \frac{3-l}{2} \quad \text{or} \quad \lambda = \frac{4+l}{2} \quad (4.31)$$

by starting from classical solutions of matrix model obtained with more general ansatze:

$$X_i = A_i^{j_1,j_2,..,j_{l-1}} L_{j_1}...L_{j_{l-1}}. \quad (4.32)$$

The case with $l = 0$ leading to a constant scalar field can be easily discussed starting with the ansatz

$$X_i = f(\rho)L_i = \frac{f(\rho)}{\rho}\hat{x}_i \quad (4.33)$$

i.e. a deformation with a scale factor of the standard fuzzy sphere solution. By imposing the equations of motion (4.1), one obtains the following quadratic form in $f(\rho)$:

$$f^2(\rho) - \lambda\rho f(\rho) - \rho^2(1-\lambda) = 0 \tag{4.34}$$

whose solutions are

$$\begin{aligned} f_1(\rho) &= \rho \quad \text{standard fuzzy sphere} \\ f_2(\rho) &= (\lambda - 1)\rho \quad \text{rescaled fuzzy sphere.} \end{aligned} \tag{4.35}$$

It is worth to notice that for $\lambda = 2$ the two solutions coincide. That this is a special point is confirmed by evaluating the classical action $S(\lambda)$ around $\lambda = 2$

$$\begin{aligned} S(\lambda)|_2 &= \frac{\rho^4}{2g^2}(\lambda-1)^3(1-\frac{\lambda}{3})Tr(\hat{L}_i\hat{L}_i) \\ S(\lambda)|_{\text{fuzzy sphere}} &= -\frac{\rho^4}{2g^2}(\frac{1}{2}-\frac{\lambda}{3})Tr(\hat{L}_i\hat{L}_i) \\ S(\lambda)|_2 - S(\lambda)|_{\text{fuzzy sphere}} &= -\frac{\rho^4}{6g^2}\lambda(\lambda-2)^3 Tr(\hat{L}_i\hat{L}_i). \end{aligned} \tag{4.36}$$

which goes as $(\lambda-2)^3$. Exactly in the same way as before, we can build, by expanding $\lambda = 2 + \rho\lambda_0$ noncommutative solitons which tend to a constant scalar field $\phi = c$ and null field strength $F_{ab} = 0$, in the classical limit.

It is possible also to verify that not only the noncommutative solution X_i tends smoothly to the classical soliton (4.27), (4.28) and (4.29), but also that the action $S(\lambda)$ evaluated at the solution X_i has the same limiting value of the classical action (3.22) evaluated on the classical soliton on the sphere.

Recall that the computation of the action $S(\lambda)$ on the general solution X_i (4.8) is given by:

$$S(\lambda)|_{tot} = S_0 + \lambda S_1 = -\frac{1}{3g^2}\rho^4(1-\lambda)(2-2\lambda+4\lambda^2)Tr(\hat{L}_m\hat{L}_m). \tag{4.37}$$

This action contains a part which is divergent in the $N \to \infty$ limit, coinciding with the value on the fuzzy sphere background, and the remaining part which is finite. Subtracting the background contribution we find

$$S(\lambda)|_{finite} = S(\lambda)|_{tot} - S(\lambda)|_{div} = \frac{4\rho^4}{3g^2}(\lambda - \frac{1}{2})^3 Tr(\hat{L}_i\hat{L}_i). \tag{4.38}$$

By introducing the models $\lambda = \frac{1}{2} + \rho\lambda_0$, considered in this paper, we find that

$$S(\lambda)|_{finite} = \frac{4}{3g^2}\rho^7\lambda_0^3 Tr(\hat{L}_i\hat{L}_i) = \frac{16\pi\lambda_0^3}{3R^2 g_{ym}^2}\rho \to 0 \quad \text{for} \quad N \to \infty \tag{4.39}$$

i.e. the finite value of the action $S(\lambda)$ vanishes as ρ in the $N \to \infty$ limit.

Before comparing this null result with the value of the classical action at the classical soliton, we must compute the possible finite contribution of the subleading terms, listed in Eq. (4.23), into the fluctuations, i.e. terms of order ρ with respect to the classical solitons, to the action $S(\lambda)$.

If we take a separate term into the action, for example the term

$$Tr[X_i, X_j][X_i, X_j] = -\frac{8}{3}\rho^4(1-\lambda)(4\lambda^2 - \lambda + 1)Tr(\hat{L}_i\hat{L}_)). \tag{4.40}$$

there is a residual finite contribution to the action coming from the subleading terms. By substituting the value $\lambda = \frac{1}{2} + \rho\lambda_0$ in Eq. (4.40), the total contribution, divergent plus finite, is given by

$$Tr[X_i, X_j][X_i, X_j]|_{total} = -\frac{8}{3}\rho^4(\frac{3}{4} - \rho^2\lambda_0^2)Tr(\hat{L}_i\hat{L}_i). \tag{4.41}$$

to be compared with the contribution of the leading fluctuations a_i, defining the classical solitons, which is different

$$Tr[X_i, X_j][X_i, X_j]|_{soliton} = -\frac{8}{3}\rho^4(\frac{3}{4} + \rho^2\lambda_0^2)Tr(\hat{L}_i\hat{L}_i). \tag{4.42}$$

It is clear that the difference is given by the subleading fluctuations $\tilde{a}_i \sim O(\rho)$, which vanish in the classical limit,

$$Tr[X_i, X_j][X_i, X_j]|_{sub} = \frac{16}{3}\rho^6\lambda_0^2 Tr(\hat{L}_i\hat{L}_i). \tag{4.43}$$

With an analogous reasoning the other two pieces of action, evaluated on the total solution (4.23), on the leading fluctuations (4.24) and on the subleading give rise to

$$\begin{aligned}
&\epsilon^{ijk}TrX_i[X_j, X_k]|_{total} = 2i\rho^3(1 - 4\lambda_0^2\rho^2)Tr(\hat{L}_i\hat{L}_i) \\
&\epsilon^{ijk}TrX_i[X_j, X_k]|_{soliton} = 2i\rho^3(1 - 2\lambda_0^2\rho^2)Tr(\hat{L}_i\hat{L}_i) \\
&\epsilon^{ijk}TrX_i[X_j, X_k]|_{sub} = -4i\lambda_0^2\rho^5 Tr(\hat{L}_i\hat{L}_i) \\
&TrX_iX_i|_{total} = \frac{4}{3}\rho^2[\frac{3}{4} + \lambda_0^2\rho^2]Tr(\hat{L}_i\hat{L}_i) \\
&TrX_iX_i|_{soliton} = \frac{4}{3}\rho^2[\frac{3}{4} + 2\lambda_0^2\rho^2]Tr(\hat{L}_i\hat{L}_i) \\
&TrX_iX_i|_{sub} = -\frac{4}{3}\rho^4\lambda_0^2 Tr(\hat{L}_i\hat{L}_i).
\end{aligned} \tag{4.44}$$

Fortunately, these finite contributions of the subleading fluctuations cancel out for the combinations S_0 and S_1 , i.e. for parts of the action in which the background is solution to the corresponding equations of motion.

$$\begin{aligned}
&S_0 \propto (\frac{1}{4}\frac{16}{3}\rho^6 - \frac{4}{3}\rho^6)\lambda_0^2 Tr(\hat{L}_i\hat{L}_i) = 0 \\
&S_1 \propto (\frac{4}{3}\rho^6 - \frac{4}{3}\rho^6)\lambda_0^2 Tr(\hat{L}_i\hat{L}_i) = 0.
\end{aligned} \tag{4.45}$$

Therefore by considering that the value of the total action $S(\lambda)$, on the complete solution X_i, is vanishing in the $N \to \infty$ limit, and that the subleading terms give no extra contributions , we conclude that the value of the classical action on the classical solitons must be zero.

To check this for the classical action (3.22) it is enough to observe that the equation of motion

$$F_{ab} + 3i\epsilon_{ab}\sqrt{g}\phi = 0 \tag{4.46}$$

cancels the quadratic part of the action, while the other equation

$$\frac{1}{\sqrt{g}}\partial_a(\sqrt{g}g^{ab}\partial_a)\phi = 6\phi \tag{4.47}$$

cancels also the other part, since the action is homogeneously quadratic in ϕ, therefore

$$S\left(\frac{1}{2}\right)|_{soliton} = 0. \tag{4.48}$$

This property does not mean that the solution is trivial, since F_{ab} cannot be made equivalent to zero with a gauge transformation.

5 Topological Solutions

Till now we have succeeded to identify noncommutative generalizations of classical solitons but we missed the more interesting case of topological solutions. In order to define noncommutative generalization of the classical monopoles on the sphere we must appeal to another method, the construction of the finite projective modules on the fuzzy sphere, which recently we have succeeded to link with the matrix model approach [35].

We follow mainly ref. [32] which makes use of the Serre-Swan theorem. Thanks to this theorem the nontrivial connections on the two-sphere S^2 can be equivalently classified in terms of projectors (gauge invariant definition of a connection), by using the Hopf principal fibration $\pi : S^3 \to S^2$ on the two-sphere S^2, with $U(1)$ as structure group.

A non trivial connection is automatically generated with the operation of projecting a trivial module. In the case of the classical sphere the projector belongs to the algebra of $N \times N$ matrices whose entries are elements of the smooth function algebra $C^\infty(S^2)$ on the base space S^2, i.e. $M_N(C^\infty(S^2))$. The section module of the bundle on which the monopole lives can be identified with the action of a global projector $p \in M_N(C^\infty(S^2))$ on the trivial module $(C^\infty(S^2))^N$, i.e. the right module $p(C^\infty(S^2))^N$, where un element of $(C^\infty(S^2))^N$ is the vector

$$||f>>= \begin{pmatrix} f_1 \\ \cdots \\ f_N \end{pmatrix} \tag{5.1}$$

with $f_1, f_2, ..., f_N$ elements of $C^\infty(S^2)$.

To connect the projector with the standard physical connection we must be able to write it in terms of a ket-bra combination

$$p = |\psi >< \psi| \qquad < \psi| =< \psi_1, ..., \psi_N|. \tag{5.2}$$

i.e. introducing a vector valued function, which must be normalized

$$< \psi|\psi >= 1 \tag{5.3}$$

in order to assure that p is a projector since

$$p^2 = |\psi >< \psi|\psi >< \psi| = p \quad p^\dagger = p \tag{5.4}$$

There is clearly an ambiguity into the definition of the vector valued function $|\psi >$ which can be redefined up to the right action of an element $w \in U(1)$, leaving the projector p invariant.

The associated canonical connection is defined as :

$$\nabla = p \cdot d \tag{5.5}$$

and its curvature

$$\nabla^2 = p(dp)^2 = |\psi >< d\psi|d\psi >< \psi|. \tag{5.6}$$

The corresponding 1-form connection A_∇ has a very simple expression in terms of the vector-valued function $|\psi >$

$$A_\nabla =< \psi|d\psi > . \tag{5.7}$$

If one is interested in classifying all the gauge non-equivalent connections then one must explore the possible global left actions on the vector valued function $|\psi >$. Reducing this global left action to the unitary group $SU(N) = \{s|ss^\dagger = 1\}$ which preserves the normalization

$$|\psi >\rightarrow |\psi^s >= s|\psi > \qquad < \psi^s|\psi^s >= 1 \tag{5.8}$$

the connection A_∇ is left invariant

$$A^s_\nabla =< \psi^s|d\psi^s >= \psi|s^\dagger s d\psi >= A_\nabla. \tag{5.9}$$

and we remain in the same class of solutions. Gauge non equivalent connections are therefore generated with the left action of group elements not preserving the normalization.

The monopole connection, which is the nontrivial instanton on the classical sphere S^2, can be explicitly built by using the $U(1)$ principal fibration $\pi : S^3 \rightarrow S^2$. In details the fibration maps the 3-dimensional sphere:

$$S^3 = \{(z_0, z_1) \in C^2; |z_0|^2 + |z_1|^2 = 1\} \tag{5.10}$$

with right $U(1)$-action

$$S^3 \times U(1) \rightarrow S^3, \quad (z_0, z_1) \cdot w = (z_0 w, z_1 w). \tag{5.11}$$

to the two-dimensional sphere, with the following positions:

$$\begin{aligned} x_1 &= z_0\bar{z}_1 + z_1\bar{z}_0 \\ x_2 &= i(z_0\bar{z}_1 - z_1\bar{z}_0) \\ x_3 &= |z_0|^2 - |z_1|^2. \end{aligned} \tag{5.12}$$

The monopole connections of charge $-n, n \in N$ can be generated by the following vector-valued function with $N = n + 1$ components

$$< \psi_{-n}| = \left(z_0^n, ..., \sqrt{\left(\begin{array}{c} n \\ k \end{array} \right)} z_0^{n-k} z_1^k, ..., z_1^n \right) \tag{5.13}$$

where, for the general n-case we need to introduce the binomial coefficients

$$\left(\begin{array}{c} n \\ k \end{array} \right) = \frac{n!}{k!(n-k)!} \qquad k = 0, 1, ..n. \tag{5.14}$$

The normalization condition for the vector-valued function (5.13) is satisfied since the complex coordinates belong to S^3:

$$< \psi_{-n}|\psi_{-n} >= (|z_0|^2 + |z_1|^2)^n = 1. \tag{5.15}$$

The corresponding projectors are simply the ket-bra combinations:

$$p_{-n} = |\psi_{-n} >< \psi_{-n}|. \tag{5.16}$$

In this classical case all these projectors are of rank 1 since its trace is

$$Tr\, p_{-n} =< \psi_{-n}|\psi_{-n} >= 1. \tag{5.17}$$

The construction makes use of the $U(1)$ left action ambiguity intrinsic into the definition of $< \psi_{-n}|$, which is defined on the sphere S^3, while the corresponding projector is left invariant, and therefore its entries can be expressed entirely as functions of the base space S^2, as it should be.

$$\begin{aligned} < \psi_{-n}| &\rightarrow < \psi_{-n}^w| = w^n < \psi_{-n}| \qquad \forall w \in U(1) \\ p_{-n} &\rightarrow p_{-n}^w = p_{-n} \qquad \bar{w}w = 1. \end{aligned} \tag{5.18}$$

The generalization of this projector construction for monopoles of charge n is straight-forward, by considering the vector-valued function:

$$< \psi_n| = \left(\bar{z}_0^n, ..., \sqrt{\left(\begin{array}{c} n \\ k \end{array} \right)} \bar{z}_0^{n-k} \bar{z}_1^k, ..., \bar{z}_1^n \right) \tag{5.19}$$

that is again normalized to 1.

The corresponding projectors

$$p_n = |\psi_n >< \psi_n| \tag{5.20}$$

since the functions $<\psi_{-n}|$ and $<\psi_n|$ are related by complex conjugation, are related by hermiticity to p_{-n}:

$$p_n = (p_{-n})^\dagger. \tag{5.21}$$

Computing the corresponding monopoles connections is an easy task, by observing that

$$\begin{aligned} A_{\mp n} &= <\psi_{\mp n}|d\psi_{\mp n}>= \mp n(\overline{z}_0 dz_0 + \overline{z}_1 dz_1) = \mp n A_1 \\ A_1 &= \overline{z}_0 dz_0 + \overline{z}_1 dz_1. \end{aligned} \tag{5.22}$$

All these connections are proportional to the topological integer n. The integer n is related to the Chern number

$$c_1(p_{\mp n}) = -\frac{1}{2\pi i}\int_{S^2} Tr(\nabla^2_{\mp n}) = -\frac{1}{2\pi i}\int_{S^2} Tr(p_{\mp n}(dp_{\mp n})^2) = \pm n. \tag{5.23}$$

At the noncommutative level it is possible to generalize the monopole connections and projectors into corresponding entities, whose entries are elements of the fuzzy-sphere algebra [31].

Our construction [31] is based on the observation that the the Hopf principal fibration relating the classical x_i coordinates to a couple of complex coordinates can be generalized to the noncommutative case promoting the two complex coordinates to a couple of independent oscillators as in formula (2.7).

The noncommutative projectors are then constructed in terms of vector-valued operators $<\psi|$, taking values in the oscillator algebra, while the projectors will be dependent on polynomial functions of the fuzzy-sphere algebra $\hat{x}_i$ only.

To construct the projectors $p_n(\hat{x}_i)$ we have considered the $(n+1)$-dimensional vectors:

$$|\psi_n>= N_n \begin{pmatrix} (a_0)^n \\ \cdots\cdots \\ \sqrt{\begin{pmatrix} n \\ k \end{pmatrix}}(a_0)^{n-k}(a_1)^k \\ \cdots\cdots \\ (a_1)^n \end{pmatrix}. \tag{5.24}$$

with an arbitrary overall function N_n of the number operator $\hat{N}$ which is fixed by the normalization condition to be

$$\begin{aligned} &<\psi_n|\psi_n>= 1 \\ &N_n = N_n(\hat{N}) = \frac{1}{\sqrt{\prod_{i=0}^{n-1}(\hat{N}-i+n)}}. \end{aligned} \tag{5.25}$$

Then the n-monopoles projector

$$p_n = |\psi_n >< \psi_n| \tag{5.26}$$

satisfies the basic properties of a projector, due to the normalization condition (5.25). Moreover in the ket-bra combination (5.26) there appear only combinations of oscillators, commuting with the number operator $\hat{N}$ and belonging to the fuzzy-sphere algebra.

Since the fuzzy sphere algebra is finite dimensional we expect that the rank of this projector is a finite positive number. The explicit calculation gives rise to

$$Tr\ p_n = Tr\ |\psi_n >< \psi_n| = \frac{N+n+1}{N+1}\ TrI = N+n+1 < \ Tr\ I = (N+1)(n+1). \tag{5.27}$$

which is always a positive integer and less than the trace of the identity.

The noncommutative analogue of the negative charge projectors $p_{-n}(n > 0)$ is obtained by taking the adjoint of the components of the vector (5.24), apart the normalization factor, which has to be recalculated in this case.

Consider the $(n+1)$-dimensional vectors:

$$|\psi_{-n} >= N_n \begin{pmatrix} (a_0^\dagger)^n \\ \cdots\cdots \\ \sqrt{\begin{pmatrix} n \\ k \end{pmatrix}}(a_0^\dagger)^{n-k}(a_1^\dagger)^k \\ \cdots\cdots \\ (a_1^\dagger)^n \end{pmatrix}. \tag{5.28}$$

The normalization condition for the vector-valued function $< \psi_{-n}|$ fixes the function N_n to be dependent only on the number operator:

$$\begin{aligned} &< \psi_{-n}|\psi_{-n} >= 1 \\ &N_n = N_n(\hat{N}) = \frac{1}{\sqrt{\prod_{i=0}^{n-1}(\hat{N}+i+2-n)}} \\ &p_{-n} = |\psi_{-n} >< \psi_{-n}|. \end{aligned} \tag{5.29}$$

Since the normalization condition for this case is different from the positive charge case, the corresponding projectors p_{-n} are not related by a hermitian operation to the p_n cases. In fact the projector p_{-n} doesn't exist for all values of n, but only for $0 < n < N+2$, as it is clear from the normalization formula (5.29). The corresponding trace

$$Tr\ p_{-n} = Tr\ |\psi_{-n} >< \psi_{-n}| = \frac{N+1-n}{N+1}\ Tr\ I = N+1-n < \ Tr\ I. \tag{5.30}$$

is positive definite if and only if the following bound is respected:

$$n < N+1. \tag{5.31}$$

which implies automatically that in the special case $n = N + 1$ we simply obtain a null projector.

In summary it is possible to extend the classical projectors $p_{\pm n}(S^2)$ to projectors $p_{\pm n}(\hat{x}_i)$, having as entries the fuzzy sphere algebra, and this deformation is completely smooth.

To prove the existence of noncommutative monopoles tending for $N \to \infty$ to the classical ones, we must be careful since in the construction of the corresponding connections we need to introduce a derivative action. Since in the noncommutative case the only derivative operator we have at disposition is the adjoint action of the Lie derivative, we must be sure that the action of the Lie derivative $[L_i, .]$ on the vectors $|\psi_{\pm n} >$ has as a smooth limit the classical Lie derivative on the sphere, and this is not obvious in our construction, since the vectors are dependent on the oscillator algebra, which is more general than the fuzzy sphere algebra.

A preliminary check is to determine if the monopole connection, constructed from the classical projectors, satisfies the Yang-Mills equations of motion. The most convenient representation of Yang-Mills action is in terms of a classical matrix model, i.e. repeating the construction of section 3 in terms of pure classical variables.

Let us define the matrix variable

$$X_i^0 = L_i + A_i \qquad [L_i, .] = -ik_i^a \partial_a \tag{5.32}$$

where A_i is related to the Yang-Mills connection $b_a(\Omega)$ and to an auxiliary scalar field $\phi(\Omega)$ as follows:

$$A_i(\Omega) = k_i^a b_a(\Omega) + \frac{x_i}{R}\phi(\Omega) \qquad \Omega = (\theta, \phi) \tag{5.33}$$

and k_i^a are the Killing vectors on the sphere S^2, and R is its radius.

The classical Yang-Mills action can be recast in terms of a matrix model action, whose dependence is polynomial in the fundamental variables:

$$\begin{aligned} S(\lambda) &= S_0 + \lambda S_1 = -\frac{1}{g_{YM}^2}\int d\Omega \left[\frac{1}{4}[X_i^0, X_j^0][X_i^0, X_j^0] - \frac{2}{3}i\lambda\epsilon^{ijk}X_i^0X_j^0X_k^0\right. \\ &+ \left.(1-\lambda)X_i^0X_i^0\right]. \end{aligned} \tag{5.34}$$

It is not difficult to link a generic projector p_0 to the matrix variables X_i^0 according to the following formula:

$$\tilde{X}_i = p_0 L_i p_0. \tag{5.35}$$

The matrix variable $\tilde{X}_i$ cannot be directly identified with X_i^0, since the projector p_0 is a gauge invariant definition of a connection. However deconstructing the projector p_0 in the form of a ket-bra valued function:

$$p_0 = |\psi>< \psi| \tag{5.36}$$

the passage from $\tilde{X}_i$ to X_i^0 is manifest:

$$X_i^0 =< \psi|\tilde{X}_i|\psi >=< \psi|L_i|\psi >= L_i+ < \psi|[L_i, |\psi >]. \tag{5.37}$$

As a byproduct we have found a representation of the connection A_i in terms of the vector-valued function $|\psi >$:

$$A_i =< \psi|[L_i, |\psi >]. \tag{5.38}$$

Introducing directly the link (5.35) between matrix model variables and projectors into the classical action (5.34) is a misleading procedure, since then the equations of motion obtained by varying the projector p_0, subject to the conditions $p_0^2 = p_0$, $p_0^\dagger = p_0$, contain more solutions than those obtained by varying the connection X_i^0. To avoid such ambiguity it is preferable to introduce the link with the projectors (5.35) at the level of Yang-Mills equations of motion. The Yang-Mills equations of motion, obtained by varying the classical Yang-Mills action with respect to X_i^0, are

$$[X_j^0, F_{ij}] = i(\lambda - 1)\epsilon_{ijk}F_{jk} \tag{5.39}$$

where

$$F_{ij} = [X_i^0, X_j^0] - i\epsilon_{ijk}X_k^0. \tag{5.40}$$

We know already that at the particular value $\lambda = 2$ these matrix model equations of motion coincide precisely with those of pure Yang-Mills theory on a two-sphere, by simply posing the auxiliary scalar field $\phi = 0$.

These equations can be recast into equations of motion for a generic projector p_0. By computing F_{ij}

$$F_{ij} = p_0([L_i, p_0][L_j, p_0] - [L_j, p_0][L_i, p_0]) \tag{5.41}$$

where we made use of the property:

$$p_0[L_i, p_0]p_0 = 0. \tag{5.42}$$

the equations (4.1), rewritten in terms of a generic projector p_0, take the following form:

$$\begin{aligned} &p_0[L^j, ([L_i, p_0][L_j, p_0] - [L_j, p_0][L_i, p_0])]p_0 = \\ &= i(\lambda - 1)\epsilon_{ijk}p_0([L_i, p_0][L_j, p_0] - [L_j, p_0][L_i, p_0])p_0. \end{aligned} \tag{5.43}$$

These equations are solved by the monopole projectors (5.16), (5.20) as we can verify directly for the simplest case, the monopole with charge 1:

$$p_1 = \frac{1}{2}\begin{pmatrix} 1 + cos\theta & sin\theta\, e^{-i\phi} \\ sin\theta\, e^{i\phi} & 1 - cos\theta \end{pmatrix}. \tag{5.44}$$

In fact if we compute the curvature we get

$$\begin{aligned} F_{\theta\phi} &= p_1(\partial_\theta p_1 \partial_\phi p_1 - \partial_\phi p_1 \partial_\theta p_1) = \frac{i}{2} sin\theta \, p_1 \\ F_{ij} &= -(k_i^\theta k_j^\phi - k_i^\phi k_j^\theta) F_{\theta\phi} = -\frac{i}{2} p_1 \epsilon_{ijk} \frac{x^k}{R}. \end{aligned} \tag{5.45}$$

Verifying that p_1 is solution of the equations of motion (5.43) for $\lambda = 2$ reduces to an algebraic question:

$$[L_j, x_k] = i\epsilon_{jkl} x^l \tag{5.46}$$

At the noncommutative level we can verify that the noncommutative projectors p_n we have constructed previously are again solutions to the equations of motion of the matrix model (5.43). Using the same link between projectors and matrix model variables as in eq. (5.35), and the Lie derivative L_i defined in terms of the oscillator algebra:

$$\begin{aligned} L_x &= \frac{1}{2}(a_0 a_1^\dagger + a_1 a_0^\dagger) \\ L_y &= \frac{1}{2}[i(a_0 a_1^\dagger - a_1 a_0^\dagger)] \\ L_z &= \frac{1}{2}(a_0 a_0^\dagger - a_1 a_1^\dagger) \end{aligned} \tag{5.47}$$

we easily obtain, using the oscillator algebra, *

$$\begin{aligned} &\tilde{X}_i^{(n)} = \\ &p_n L_i p_n = \frac{N}{N+n} |\psi_n > L_i < \psi_n| = |\psi_n > L_i < \psi_n| + |\psi_n > A_i^{(n)} < \psi_n| \;\; n > 0 \\ &\tilde{X}_i^{(-n)} = \\ &p_{(-n)} L_i p_{(-n)} = \frac{N+2}{N+2-n} |\psi_{(-n)} > L_i < \psi_{(-n)}| = |\psi_{(-n)} > L_i < \psi_{(-n)}| + \\ &+ |\psi_{(-n)} > A_i^{(-n)} < \psi_{(-n)}| \quad 0 < n < N+1. \end{aligned} \tag{5.48}$$

At finite N, a nonlinear dependence on n in the denominator appears in the (gauge invariant) connection

$$\begin{aligned} |\psi_n > A_i^{(n)} < \psi_n| &= -\frac{n}{N+n} |\psi_n > L_i < \psi_n| \;\; n > 0 \\ |\psi_{(-n)} > A_i^{(-n)} < \psi_{(-n)}| &= \frac{n}{N+2-n} |\psi_{(-n)} > L_i < \psi_{(-n)}| \\ & \quad 0 < n < N+1. \end{aligned} \tag{5.49}$$

*For the simplest case $n = 1, N = 1$ the result can be expressed as a sum of two independent angular momenta, one is intrinsic to the fuzzy sphere algebra $L_i \otimes 1$ and the other $1 \otimes S_i$ acting on the $(2 \otimes 2)$ auxiliary representation space of the projector.

The same is true for the field strength:

$$
\begin{aligned}
\tilde{F}^{(n)}_{ij} &= -\frac{nN}{(N+n)^2} i\epsilon_{ijk}|\psi_n > L_k < \psi_n| \quad n > 0 \\
\tilde{F}^{(-n)}_{ij} &= \frac{n(N+2)}{(N+2-n)^2} i\epsilon_{ijk}|\psi_{(-n)} > L_k < \psi_{(-n)}| \\
& \quad 0 < n < N+1.
\end{aligned} \tag{5.50}
$$

The corresponding equations of motion are solved by a certain value of λ, due to the identity:

$$
\begin{aligned}
&[\tilde{X}^{(n)}_j, \tilde{F}^{(n)}_{ij}] = i\frac{N}{N+n}\epsilon_{ijk}\tilde{F}^{(n)}_{jk} \quad \lambda = 2 - \frac{n}{N+n} \quad n > 0 \\
&[\tilde{X}^{(-n)}_j, \tilde{F}^{(-n)}_{ij}] = i\frac{N+2}{N+2-n}\epsilon_{ijk}\tilde{F}^{(-n)}_{jk} \quad \lambda = 2 + \frac{n}{N+2-n} \quad 0 < n < N+1
\end{aligned} \tag{5.51}
$$

There is a class of models situated around the classical value $\lambda = 2$ admitting noncommutative extensions of topological solutions analogously to the case of noncommutative soliton solutions. The condition that the noncommutative solutions are smooth extensions of the classical ones always requires to perturb the λ coupling constant in a similar form:

$$
\lambda = \lambda_{cl} + \frac{c}{N}. \tag{5.52}
$$

The special case $\lambda_{cl} = 2$ deserves an explanation, since soliton and topological solutions are interconnected. In section 4 we found a class of solutions tending to the classical solution $\phi = \text{const.}$, $b_a = 0$, in which case the matrix model variable X_i was equal to a rescaling of the background solution by a factor $(1 + f(\frac{1}{N}))$:

$$
X_i = (1 + f(\frac{1}{N}))\hat{x}_i \quad \rightarrow \quad \phi = \text{const} \quad b_a = 0. \tag{5.53}
$$

Evidently, inside the class of models (5.51) it is possible to reach a different classical topological solution

$$
X_i \rightarrow \phi = 0, \quad b_a = \text{monopole} \tag{5.54}
$$

In the following we will construct the noncommutative map relating the two class of solutions. This requires another step, i.e. reconstructing the monopole connections from the projectors $p_{\pm n}$.

At a classical level this reconstruction is an easy task. Let us recall the formula (5.38)

$$
A_i = < \psi|[L_i, |\psi >] \tag{5.55}
$$

and let us try to apply it to the case of the simplest monopole $n = 1$

$$
< \psi_1| = (\overline{z}_0, \overline{z}_1) \tag{5.56}
$$

with

$$z_0 = cos\frac{\theta}{2}e^{i(\frac{\psi-\phi}{2})} \quad z_1 = sin\frac{\theta}{2}e^{i(\frac{\psi+\phi}{2})} \tag{5.57}$$

The commutator action of the Lie derivative reduces to a derivative action in the classical case:

$$\begin{aligned} [L_i, .] &= -i(k_i^\theta \partial_\theta + k_i^\phi \partial_\phi + k_i^\psi \partial_\psi) \\ k_i^\theta &= (-sin\phi, cos\phi, 0) \\ k_i^\phi &= (-cos\phi cotg\theta, -sin\phi cotg\theta, 1) \\ k_i^\psi &= (-\frac{cos\phi}{sin\theta}, -\frac{sin\phi}{sin\theta}, 0) \end{aligned} \tag{5.58}$$

where k_i^θ, k_i^ϕ are the killing vector of the sphere and k_i^ψ is the residue of the extended action on S^3. Since the vector-valued function $< \psi|$ depends on the total space S^3, the derivative action contains the dependence on the auxiliary $U(1)$ variable ψ of the three sphere S^3. This ruins the classical limit, since we must require that the Lie derivative action is equivalent to the ordinary derivate action $d = d\theta\frac{\partial}{\partial\theta} + d\phi\frac{\partial}{\partial\phi}$. It is not difficult to convince ourself that this is possible if and only if the vector-valued function $< \psi_n|$ can be projected to the basic space S^2, keeping the projectors invariant. It turns out that the presence of the spurious variable ψ can be removed by redefining the vector-valued function $< \psi_1|$:

$$< \psi_1| \rightarrow < \psi_1'| = e^{i\frac{\psi}{2}} < \psi_1| \quad \text{for } n = 1 \tag{5.59}$$

or in the general case

$$< \psi_n| \rightarrow < \psi_n'| = e^{in\frac{\psi}{2}} < \psi_n| \quad \forall\, n \in N. \tag{5.60}$$

After this redefinition, which doesn't alter the projectors, we obtain the well-known result for the monopole connection:

$$b_\theta = 0 \;\; b_\phi = -\frac{n}{2}cos\theta. \tag{5.61}$$

This redefinition, which is simple in the classical case, turns out to be necessary also in the noncommutative case. In fact the vector $< \psi_n(\hat{x}_i)|$, used for the computation in (5.24), is function of the oscillator algebra, which is more general of the fuzzy sphere algebra. This represents a problem when acting with the Lie derivative, since its classical limit doesn't reduce to the usual action of the Lie derivative on the sphere S^2.

At the classical level we solved the problem using the $U(1)$ gauge arbitrariness $|\psi > \rightarrow |\psi > e^{i\phi}$ which leaves the projector invariant.

The analogue of this phase ambiguity at a noncommutative level is suggested by looking at the explicit form of the vector-valued operator $|\psi >$, for example in the case $n = -1$:

$$|\psi_{n=-1} > \;=\; \frac{1}{\sqrt{\hat{N}+1}}\begin{pmatrix} \bar{a}_0 \\ \bar{a}_1 \end{pmatrix} = \begin{pmatrix} \tilde{\bar{a}}_0 \\ \tilde{\bar{a}}_1 \end{pmatrix}$$

$$
\begin{aligned}
\bar{\tilde{a}}_0 &= \sum_{n_1=0}^{\infty}\sum_{n_2=0}^{\infty}\sqrt{\frac{n_1+1}{n_1+n_2+2}}|n_1+1,n_2><n_1,n_2| \\
\bar{\tilde{a}}_1 &= \sum_{n_1=0}^{\infty}\sum_{n_2=0}^{\infty}\sqrt{\frac{n_2+1}{n_1+n_2+2}}|n_1,n_2+1><n_1,n_2| \\
<\psi_{n=-1}| &= (a_0,a_1)\frac{1}{\sqrt{\hat{N}+1}} = (\tilde{a}_0,\tilde{a}_1) \\
\tilde{a}_0 &= \sum_{n_1=0}^{\infty}\sum_{n_2=0}^{\infty}\sqrt{\frac{n_1+1}{n_1+n_2+2}}|n_1,n_2><n_1+1,n_2| \\
\tilde{a}_1 &= \sum_{n_1=0}^{\infty}\sum_{n_2=0}^{\infty}\sqrt{\frac{n_2+1}{n_1+n_2+2}}|n_1,n_2><n_1,n_2+1| \\
<\psi_{-1}|\psi_{-1}> &= 1.
\end{aligned}
\tag{5.62}
$$

The problem which we must solve is that the action of $|\psi_{-1}>$ doesn't commute with the number operator $\hat{N}$ and therefore it is not possible to restrict its action to a fixed number N, as instead we can do for the projectors.

It is then necessary to redefine the vector $|\psi_{-1}>$ with an operator, acting on the right and not commuting with the number operator, i.e. a quasi-unitary operator:

$$
|\psi_{-1}>\rightarrow|\psi'_{-1}>=|\psi_{-1}>U \qquad UU^{\dagger}=1 \quad (U^{\dagger}U=1-P_0) \tag{5.63}
$$

in order to keep invariant the noncommutative projectors. The presence of the quasi-unitary operator U is necessary to extrapolate the noncommutative connection. Many choices for U are possible, for example:

$$
U_1=\sum_{n_1=0}^{\infty}\sum_{n_2=0}^{\infty}|n_1,n_2><n_1+1,n_2| \tag{5.64}
$$

or

$$
U_2=\sum_{n_1=0}^{\infty}\sum_{n_2=0}^{\infty}|n_1,n_2><n_1,n_2+1| \tag{5.65}
$$

are equally good, since the difference between these two operators is a gauge transformation of the connection. Let's compute

$$
\begin{aligned}
|\psi'_{-1}> &= |\psi_{-1}>U_1 = \begin{pmatrix}\bar{\tilde{a'}}_0\\ \bar{\tilde{a'}}_1\end{pmatrix} \\
\bar{\tilde{a'}}_0 &= \sum_{n_1=0}^{\infty}\sum_{n_2=0}^{\infty}\sqrt{\frac{n_1+1}{n_1+n_2+2}}|n_1+1,n_2><n_1+1,n_2| \\
\bar{\tilde{a'}}_1 &= \sum_{n_1=0}^{\infty}\sum_{n_2=0}^{\infty}\sqrt{\frac{n_2+1}{n_1+n_2+2}}|n_1,n_2+1><n_1+1,n_2|.
\end{aligned}
\tag{5.66}
$$

Since now $[\hat{N}, |\psi'_{-1}>] = 0$ we can truncate the action of $|\psi'_{-1}>$ to a fixed number N:

$$\begin{aligned}\tilde{a'}_0|_N &= \sum_{k=0}^{N-1}\sqrt{\frac{k+1}{N+1}}|k+1, N-k-1><k+1, N-k-1| \\ \tilde{a'}_1|_N &= \sum_{k=0}^{N-1}\sqrt{\frac{N-k}{N+1}}|k, N-k><k+1, N-k-1|.\end{aligned} \tag{5.67}$$

These actions belong to the physical functional space of the noncommutative case, and can be used to extrapolate the noncommutative connections. Since the projectors are invariant under this redefinition, it is clear that $|\psi'_{-1}>$ gives rise to connections satisfying the Y-M equations of motion (in the gauge-covariant formulation). Since

$$\tilde{X}_i = |\psi_{-1}> U\frac{N+2}{N+1}(U^\dagger L_i U)U^\dagger <\psi_{-1}| = |\psi'_{-1}> X_i <\psi_{-1}| \tag{5.68}$$

we can deduce that the physical matrix model variable X_i is of the form:

$$\begin{aligned}&X_i = \frac{N+2}{N+1}U^\dagger L_i U \\ &F_{ij} = [X_i, X_j] - i\epsilon_{ijk}X_k = \frac{N+2}{(N+1)^2}i\epsilon_{ijk}(U^\dagger L_j U) \\ &[X^j, F_{ij}] = i(\lambda-1)\epsilon_{ijk}F_{jk} \quad \lambda = 2 + \frac{1}{N+1}.\end{aligned} \tag{5.69}$$

The generalization of these results to the case $|\psi_{-n}>$ with n generic ($n < N+1$) is straightforward.

Therefore the noncommutative monopole configurations are characterized by mapping the solution (4.33), leading to the classical solution $\phi = const.$, $b_a = 0$, to another class, with the help of quasi-unitary operators. Their role is to generate reducible representations of the $SU(2)$ Lie algebra, recognizing the topological nature of these solutions. Nontrivial topologies on a fuzzy sphere are then classified by reducible representations of the $SU(2)$ Lie algebra †.

If one tries to repeat this construction for the case $|\psi_n>,\ n>0$, for example $n=1$:

$$\begin{aligned}|\psi_{n=1}> &= \frac{1}{\sqrt{\hat{N}+1}}\begin{pmatrix} a_0 \\ a_1\end{pmatrix} = \begin{pmatrix}\tilde{a}_0 \\ \tilde{a}_1\end{pmatrix} \\ \tilde{a}_0 &= \sum_{n_1=0}^{\infty}\sum_{n_2=0}^{\infty}\sqrt{\frac{n_1+1}{n_1+n_2+1}}|n_1, n_2><n_1+1, n_2| \\ \tilde{a}_1 &= \sum_{n_1=0}^{\infty}\sum_{n_2=0}^{\infty}\sqrt{\frac{n_2+1}{n_1+n_2+1}}|n_1, n_2><n_1, n_2+1|\end{aligned}$$

†Since there are no one-dimensional representations of $SU(2)$ Lie algebra, the case $n = N$ leads to a vanishing matrix variable, also if the corresponding projector is non-vanishing.

$$
\begin{aligned}
< \psi_{n=1}| &= \frac{1}{\sqrt{\hat{N}+1}}(\bar{a}_0, \bar{a}_1) = (\tilde{\bar{a}}_0, \tilde{\bar{a}}_1) \\
\tilde{\bar{a}}_0 &= \sum_{n_1=0}^{\infty}\sum_{n_2=0}^{\infty} \sqrt{\frac{n_1+1}{n_1+n_2+1}} |n_1+1, n_2 >< n_1, n_2| \\
\tilde{\bar{a}}_1 &= \sum_{n_1=0}^{\infty}\sum_{n_2=0}^{\infty} \sqrt{\frac{n_2+1}{n_1+n_2+1}} |n_1, n_2+1 >< n_1, n_2|.
\end{aligned} \tag{5.70}
$$

we are forced to work with the adjoint of the quasi-unitary operator U

$$
U^{\dagger} = \sum_{n_1=0}^{\infty}\sum_{n_2=0}^{\infty} |n_1+1, n_2 >< n_1, n_2|. \tag{5.71}
$$

with the result of altering the form of the projector p_1. We have found that it is impossible for the charge n monopoles to define a connection satisfying the Y-M equations of motion, while the corresponding projector does it. There is an simple argument in agreement with this picture, since in a $(N+1) \times (N+1)$ matrix variable X_i we can insert a (reducible) representation with rank less than $(N+1)$, but not bigger than $(N+1)$.

We conclude that at noncommutative level there is in general no equivalence relation between projectors and connections, as it happens in the classical case.

6 Conclusions

This review summarizes how to introduce noncommutative configurations like solitons or monopoles on a fuzzy sphere which smoothly extend the commutative ones following Refs. [27] and [35]. Instead than trying to solve the equations of motion of the noncommutative gauge theory directly we can define the noncommutative soliton as a difference between a generic solution of the matrix model and the fuzzy sphere background, when this difference satisfies the prerequisite of good fluctuation, i.e. it vanishes as $\frac{1}{N}$ in the $N \to \infty$. In order to achieve this goal we have to choose an ansatz, for example

$$
X_i = A_i^j L_j \tag{6.1}
$$

and correspondingly we perturb the coupling constant λ as $\lambda = \frac{1}{2} + \frac{\lambda_0}{N}$. At fixed N the corresponding class of solutions determines a series of noncommutative solitons on the fuzzy sphere, converging to a classical soliton in the $N \to \infty$ limit, where the scalar field has a fixed angular momentum $l = 2$. By analyzing the classical limit of these solitons we find confirmations of the scheme proposed.

Noncommutative extensions of classical solitons with higher spherical harmonics $l > 2$ are probably generated by the same method, starting from a general ansatz:

$$
X_i = A_i^{j_1 j_2 \cdots j_{l-1}} L_{j_1 j_2 \cdots j_{l-1}} \tag{6.2}
$$

The special case $\lambda = 2$, corresponding to the scalar case $l = 0$ has been studied carefully. In this case soliton and topologically nontrivial configurations are interconnected since in the classical limit are both solutions of the same matrix model. In the noncommutative case we find that there exist noncommutative extensions of both solutions, and that a simple map, based on the quasi unitary operators, relates them.

The type of ansatz (6.2) unfortunately truncates the maximum angular momentum to a fixed value, thus preventing from the possibility of extrapolating the noncommutative monopoles, which contain all the values of the angular momentum in the $N \to \infty$ limit, and they cannot be reached by this method.

To introduce topologically nontrivial configurations on the fuzzy sphere we must appeal to the method of projective modules, which already in the classical case describe correctly the monopole solution. We use the results of ref. [31] , where we were able to construct some noncommutative projectors having as entries the elements of the fuzzy sphere algebra, and tending smoothly in the $N \to \infty$ limit to the classical monopoles on the sphere S^2.

By introducing a simple link between projectors and matrix models we firstly have succeeded to identify a class of models for which the noncommutative projectors [31] are solutions of the corresponding equations of motion. However the more interesting question was to understand the structure of the connections or equivalently the matrix model variables X_i corresponding to these projectors. This requires to be able to deconstruct the projectors in terms of more fundamental vector-valued fields, and this is possible only for projectors with negative charge $-n$ ($p_{-n}, n > 0$), but not those of positive charge $p_n, n > 0$.

In the $p_{-n}(n > 0)$ case, we have isolated the corresponding matrix model solution, which is obtained by the soliton solution corresponding to the scalar case $l = 0$, dressing with a quasi-unitary operator. The background, which is an irreducible representation of the $SU(2)$ Lie algebra, is mapped to a reducible representation of the $SU(2)$ Lie algebra, thus revealing its topological nature.

The case of positive charge projectors p_k $(k > 0)$, where it is not possible to define a corresponding connection satisfying the $Y - M$ equations of motion is unexpected from the point of view of the theorem of Serre-Swan which assures the complete equivalence of projective modules and connections at a classical level.

References

[1] E. Witten, " *Noncommutative geometry and string field theory* ", *Nucl. Phys.* **B268** (1986) 253.

[2] N. Seiberg and E. Witten, " *String Theory and Noncommutative geometry* ", *JHEP* **9909** (1999) 032, hep-th/9908142.

[3] A. Connes, M. R. Douglas and A. Schwarz, " *Noncommutative geometry and Matrix theory: compactification on Tori* ", *JHEP* **9802** (1998) 003, hep-th/9711162.

[4] J. Madore, " *The Fuzzy sphere* ", *Class. Quantum Grav.* **9** (1992) 69.

[5] T. Masuda, " *Normalized Weyl-type *-product on Kahler manifolds* ", *Mod. Phys. Lett.* **A15** (2000) 2177, hep-th/0010009.

[6] U. Carow-Watamura and S. Watamura, " *Noncommutative geometry and gauge theory on Fuzzy sphere* ", *Comm. Math. Phys.* **212** (2000) 395, hep-th/9801195.

[7] U. Carow-Watamura and S. Watamura, " *Differential calculus on fuzzy sphere and scalar field* ", *Int. J. Mod. Phys.* **A13** (1998) 3235, q-alg/9710034.

[8] U. Carow-Watamura and S. Watamura, " *Chirality and Dirac operator on Noncommutative sphere* ". *Comm. Math. Phys.* **183** (1997) 365, hep-th/9605003.

[9] C. Klimcik, " *Gauge theories on the noncommutative sphere* ", *Comm. Math. Phys.* **199** (1998) 257, hep-th/9710153.

[10] H. Grosse and P. Presnajder, " *The Dirac operators on the fuzzy sphere* ", *Lett. Math. Phys.* **33** (1995) 171.

[11] J. Ambjorn, Y.M. Makeenko, J. Nishimura and R. J. Szabo, " *Finite N Matrix models of noncommutative gauge theory* ", *JHEP* **9911** (1999) 029, hep-th/9911041.

[12] J. Ambjorn, Y.M. Makeenko, J. Nishimura and R. J. Szabo, " *Non perturbative Dynamics of noncommutative gauge theory* ", *Phys. Lett.* **B480** (2000) 399.

[13] J. Ambjorn, Y.M. Makeenko, J. Nishimura and R. J. Szabo, " *Lattice gauge fields and discrete noncommutative Yang-Mills Theory* ", *JHEP* **0005** (2000) 023, hep-th/0004147.

[14] T. Eguchi and K. Kawai, " *Reduction of dynamical degrees of freedom in the large N gauge theory* ", *Phys. Rev. Lett.* **48** (1982) 1063.

[15] A. Gonzales-Arroyo and M. Okawa, " *The twisted Eguchi-Kawai model: a reduced model for large N lattice gauge theory* ", *Phys. Rev.* **D27** (1983) 2397.

[16] T. Banks, W. Fischler, S. Shenker and L. Susskind, " *M theory as a matrix model: a conjecture* " *Phys. Rev.* **D55** (1997) 5112, hep-th/9610043.

[17] N. Ishibashi, H. Kawai, Y. Kitazawa and A. Tsuchiya, " *A large N reduced model as superstring* ", *Nucl. Phys.* **B498** (1997) 467, hep-th/9612115.

[18] H. Aoki, S. Iso, H, Hawai, Y. Kitazawa, T. Tada and A. Tsuchiya, " *IIB matrix model* ", *Prog. Theor. Phys. Suppl.* **134** (1999) 47, hep-th/9908038.

[19] H. Aoki, S. Iso, H. Hawai, Y. Kitazawa and T. Tada, " *Space-time structures from IIB Matrix model* ", *Prog. Theor. Phys.* **99** (1998) 713, hep-th/9802085.

[20] M. Li, " *Strings from IIB matrices* ", *Nucl. Phys.* **B499**(1997) 149,hep-th/9612222.

[21] H. Aoki, N. Ishibashi, S. Iso, H. Kawai, Y. Kitazawa and T. Tada, " *Noncommutative Tang-Mills in IIB Matrix model* ", *Nucl. Phys.* **B565** (2000) 176, hep-th/9908141.

[22] S. Iso, Y. Kimura, K. Tanaka and K. Wakatsuki, " *Noncommutative gauge theory on fuzzy sphere from Matrix model* ", *Nucl. Phys.* **B604** (2001) 121, hep-th/0101102.

[23] Y. Kimura, " *Noncommutative gauge theories on Fuzzy sphere and Fuzzy torus from Matrix model* ", *Prog. Theor. Phys.* **106** (2001) 445, hep-th/0103192.

[24] R.C. Myers, *"Dielectric branes", JHEP* **9912** (1999) 022, hep-th/9910053.

[25] D. Berenstein, J. M. Maldacena, H. Nastase, " *Strings in flat space and PP waves from N=4 Superyang-mills* ", *JHEP* **0204** (2002) 013, hep-th/0202021.

[26] P. Valtancoli, *"Stability of the fuzzy sphere solution from matrix model", Int. J. Mod. Phys.* **A18** (2003) 967, hep-th/0206075.

[27] P. Valtancoli, " *Solitons for the fuzzy sphere from matrix model* ", *Int. J. Mod. Phys.* **A18** (2003) 1107, hep-th/0209117.

[28] S. Baez, A.P. Balachandran, B. Ydri, S. Vaidya, " *Monopoles and Solitons in fuzzy physics", Comm. Math. Phys.* **208** (2000) 787, hep-th/9811169.

[29] A.P. Balachandran and S. Vaidya, *"Instantons and chiral anomaly in fuzzy physics"*, hep-th/9910129.

[30] A. P. Balachandran, X.Martin and D. O' Connor, " *Fuzzy actions and their continuum limits* ", hep-th/0007030.

[31] P. Valtancoli, " *Projectors for the fuzzy sphere* ", *Mod. Phys. Lett.* **A16** (2001) 639, hep-th/0101189.

[32] G. Landi, " *Projective modules of finite type and monopoles over* S^2 ", *J. Geom. Phys.* **37** 47, 2001, math-ph/9905014.

[33] M.Dubois-Violette, Y. Georgelin, *"Gauge theory in terms of projector valued fields", Phys. Lett. B* **82** 251, 1979.

[34] T. Eguchi, P. B. Gilkey, A. J. Hanson, " *Gravitation, gauge theories and differential geometry* ", *Phys. Rept.* **66** 213,1980.

[35] P. Valtancoli, " *Projectors, matrix models and noncommutative monopoles* ",hep-th/0404045, to be published in *Int. Journal of Modern Physics.*

In: New Developments in Soliton Research
Editor: L.V. Chen pp. 87-100
ISBN 1-59554-561-8

Chapter 3

NONLINEAR DYNAMICS OF SELF-GUIDING ELECTROMAGNETIC BEAMS IN RELATIVISTIC ELECTRON-POSITRON PLASMAS

***M. Ohhashi*[1], *T. Tatsuno*[1,2], *V.I. Berezhiani*[3] *and S.V. Mikeladze*[3]**
[1]Graduate School of Frontier Sciences,
University of Tokyo, Hongo 7-3-1, Tokyo 113-0033, Japan
[2]Institute for Research in Electronics and Applied Physics,
University of Maryland, College Park, Maryland 20742-3511, USA
[3]Institute of Physics, The Georgian Academy of Sciences,
6 Tamarashvili str., Tbilisi 380077, Georgia

Abstract

Nonlinear interaction of an intense electromagnetic (EM) beam with relativistically hot electron-positron plasma is investigated by invoking the variational principle and numerical simulation, resting on the model of generalized nonlinear Schrödinger equation with saturating nonlinearity. The present analysis shows the dynamical properties including the possibilities of trapping and wave-breaking of EM beams. These properties of EM beams may give a significant clue for the gamma-ray burst.

1 Introduction

The problem of electromagnetic (EM) wave propagation and related phenomena in relativistic plasma have attracted considerable attention in the recent past. Relativistic electron-positron (e-p) dominated plasmas are created in a variety of astrophysical situations. Electron-positron pair production cascades are believed to occur in pulsar magnetospheres [1]. The e-p plasmas are also likely to be found in the bipolar outflows (Jets) in Active Galactic Nuclei (AGN) [2], and at the center of our own Galaxy [3]. In AGNs, the observations of superluminal motions are commonly attributed to the expansion of relativistic e-p beams in a pervading subrelativistic medium. This model implies copious pair production via γ-γ interactions creating an e-p atmosphere around the source. The actual production of e-p pairs due to photon-photon interactions occurs in the coronas of AGN accretion disks, which upscatter the soft photons emitted by the accretion disks by inverse Compton scattering. The presence of e-p plasma is also argued in the MeV epoch of the early Universe [4]. On the other hand the contemporary progress in the development of super-strong laser sources with intensities $I = 10^{21-23}$ $\mathrm{W/cm^2}$ has also made it possible to create relativistic e-p plasmas in the laboratory by a variety of experimental techniques [5]. Elucidation of the electromagnetic wave dynamics in a relativistic e-p plasmas will, perhaps, be an essential determinant of the radiation properties of astrophysical objects as well as of the medium exposed to the field of super-strong laser radiation.

Wave self-modulation and soliton-formation is, perhaps, one of the most interesting and significant features of the overall plasma dynamics. The existence of stable localized envelope solitons of EM radiation has been suggested as a potential mechanism for the production of micro-pulses in AGN and pulsars [6, 7, 8]. In the early Universe localized solitons are strong candidates to explain the observed inhomogeneities of the visible Universe [9, 10], and vortex soliton can be considered to play a significant role for the formation of the observed structure of the Universe [11]. The gamma-ray bursts (GRBs) and their afterglows are likely to be the result of energy dissipation from a relativistically expanding outflow [12]. A Poynting-flux driven outflow from a magnetized rotator is a promising paradigm for GRB engines and there have been various implementations of this concept [13, 14, 15, 16, 17]. Recently Lyutikov and Blackman suggested that gamma-rays are emitted at the point where MHD breaks down due to the overturn instability of large amplitude electromagnetic wave [18].

In the present paper we explore the mechanism for the localization of multi-dimensional intense EM radiation in pure e-p plasmas. Assuming the plasma to be transparent to the beam, and applying a fully relativistic hydrodynamical model, we demonstrate the possibility of beam self-trapping leading to the formation of stable 2D solitonic structures. The high-frequency pressure force of the EM field (tending to completely expel the pairs radially from the region of localization) is overwhelmed by the thermal pressure force which opposes the radial expansion of the plasma creating conditions for the formation of the stationary self-guiding regime of beam propagation.

2 Basic Equations

In this section we apply our general formulation to the problem of self-trapping of EM beams in pure e-p plasmas with relativistic temperatures. We assume that the equilibrium state of the plasma is characterized by an overall charge neutrality $n_\infty^- = n_\infty^+ \equiv n_\infty$, where n_∞^- and n_∞^+ are the unperturbed number densities of the electrons and positrons in the far region of the EM beam localization. In most mechanisms for creating e-p plasmas, the pairs appear simultaneously and due to the symmetry of the problem it is natural to assume that $T_\infty^- = T_\infty^+ \equiv T_\infty$, where T_∞^- and T_∞^+ are the respective equilibrium temperatures.

We shall assume that for the radiation field of interest, the plasma is underdense and transparent, i.e., $\epsilon = \omega_e/\omega \ll 1$, where ω is the mean frequency of EM radiation and $\omega_e = (4\pi e^2 n_\infty/m_{0e})^{1/2}$ is the plasma frequency. Following Ref. [19] we introduce a temperature dependent momentum $\mathbf{\Pi}^\pm = G^\pm \mathbf{p}^\pm$ and relativistic factor $\Gamma^\pm = G^\pm \gamma^\pm$, where $G = K_3(z)/K_2(z)$ is the temperature dependent factor with K_n the n-th order modified Bessel function of the second kind ($z = m_{0e}c^2/T$), $\mathbf{p}^\pm$ and $\gamma^\pm = \sqrt{1 + (\mathbf{p}^\pm/m_e c)^2}$ are

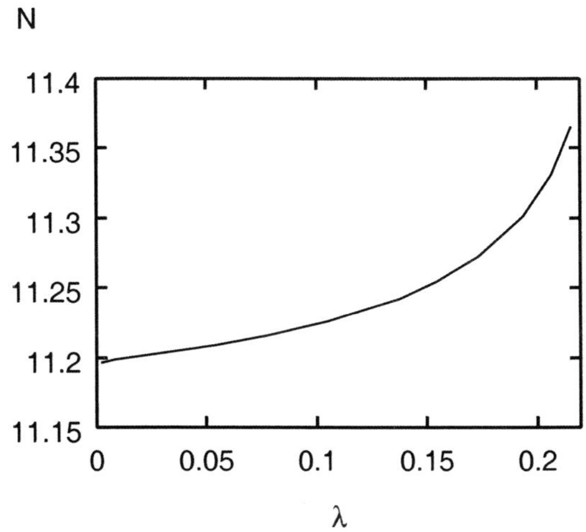

Figure 1: The beam power N versus λ ($T_\infty = 1$).

respectively momentum and relativistic factor of e-p particles. Introducing the dimensionless quantities $\tilde{t} = \omega t$, $\tilde{\mathbf{r}} = (\omega/c)\mathbf{r}$, $\tilde{T}^{\pm} = T^{\pm}/m_{0e}c^2$, $\tilde{\mathbf{A}} = e\mathbf{A}/(m_{0e}c^2)$, $\tilde{\phi} = e\phi/m_{0e}c^2$, $\tilde{\mathbf{\Pi}}^{\pm} = \mathbf{\Pi}^{\pm}/(m_{0e}c)$, and $\tilde{n}^{\pm} = n^{\pm}/n_\infty$, we arrive at the dimensionless equations [19],

$$\frac{\partial \mathbf{\Pi}^{\pm}}{\partial t} + \nabla\Gamma^{\pm} = \mp\frac{\partial \mathbf{A}}{\partial t} \mp \nabla\phi, \tag{1}$$

$$\frac{n^{\pm}}{\Gamma^{\pm} f(T^{\pm})} = \text{const}, \tag{2}$$

$$\frac{\partial n^{\pm}}{\partial t} + \nabla\cdot\mathbf{J}^{\pm} = 0, \tag{3}$$

$$\Delta\phi = \epsilon^2(n^- - n^+), \tag{4}$$

$$\frac{\partial^2 \mathbf{A}}{\partial t^2} - \Delta\mathbf{A} + \frac{\partial}{\partial t}\nabla\phi - \epsilon^2(\mathbf{J}^+ - \mathbf{J}^-) = 0, \tag{5}$$

where

$$f(T^{\pm}) = \frac{K_2(1/T^{\pm})T^{\pm}}{G^{\pm}(T^{\pm})}\exp[G^{\pm}(T^{\pm})/T^{\pm}], \tag{6}$$

with $\mathbf{J}^{\pm} = n^{\pm}\mathbf{\Pi}^{\pm}/\Gamma^{\pm}$ and $\Gamma^{\pm} = \sqrt{(G^{\pm})^2 + (\mathbf{\Pi}^{\pm})^2}$. Here $\mathbf{A}$ and ϕ are the vector and scalar potentials ($\nabla\cdot\mathbf{A} = 0$). The tilde is suppressed for convenience.

Of various techniques that could be invoked to investigate Eqs. (1)-(6) to study the self-trapping of high-frequency EM radiation propagating along the z-axis, we choose the method presented in the excellent paper by Sun *et al.* [20]. The method is based on the multiple scale expansion of the equations in the small parameter ϵ. Assuming that all variations are slow compared to the variation in $\xi = z - at$, we expand all quantities $Q = (\mathbf{A}, \phi, \mathbf{\Pi}^{\pm}, n^{\pm}, \ldots)$ as

$$Q = Q_0(\xi, x_1, y_1, z_2) + \epsilon Q_1(\xi, x_1, y_1, z_2), \tag{7}$$

where $(x_1, y_1, z_2) = (\epsilon x, \epsilon y, \epsilon^2 z)$ denote the directions of slow change, and $a_1 = (a^2 - 1)/\epsilon^2 \sim 1$. We further assume that the high-frequency EM field is circularly polarized,

$$\mathbf{A}_{0\perp} = \frac{1}{2}(\hat{\mathbf{x}} + i\hat{\mathbf{y}})A\exp(i\xi/a) + \text{c.c.}, \tag{8}$$

where A is the slowly varying envelope of the EM beam, $\hat{\mathbf{x}}$ and $\hat{\mathbf{y}}$ denote unit vectors, and c.c. is the complex conjugate.

We now give a short summary of the steps in the standard multiple-scale methodology (Ref. [20]). In the lowest order in ϵ, the transverse (to the direction of EM wave propagation z) component of Eq. (1) reduces to

$$\mathbf{\Pi}^{\pm}_{0\perp} = \mp\mathbf{A}_{0\perp}. \tag{9}$$

To the next order (in ϵ), the transverse component of Eq. (1) reads:

$$-a\frac{\partial\mathbf{\Pi}^{\pm}_1}{\partial\xi} + \nabla_\perp\Gamma^{\pm}_0 = \pm a\frac{\partial\mathbf{A}_1}{\partial\xi} \mp \nabla_\perp\phi_0. \tag{10}$$

Averaging Eq. (10) over the fast variable ξ we obtain $\nabla_\perp\Gamma^{\pm}_0 = \mp\nabla_\perp\phi_0$. Using these relations, we obtain $\nabla_\perp\phi_0 = 0$ and

$$\Gamma_0 \equiv \Gamma^{\pm}_0 = \text{const}. \tag{11}$$

Note that from the lowest order of Eqs. (2) and (4), we conclude that $n^+_0 = n^-_0 \equiv n_0$ and $T^+_0 = T^-_0 \equiv T_0$. The relation between EM field and temperature can be found by invoking Eq. (11). Using Eqs. (8) and (9) and determining integration constant such that $A \to 0$ and $T_0 \to T_\infty$, we obtain

$$G^2(T_0) = G^2(T_\infty) - |A|^2. \tag{12}$$

The equation for slowly varying envelope A of EM beam can be obtained from Eq. (5). To the lowest order we find

$$a_1\frac{\partial^2\mathbf{A}_{0\perp}}{\partial\xi^2} - \nabla^2_\perp\mathbf{A}_{0\perp} - 2\frac{\partial^2\mathbf{A}_{0\perp}}{\partial\xi\partial z_2} + 2\frac{n_0(T_0)}{G(T_\infty)}\mathbf{A}_{0\perp} = 0. \tag{13}$$

In deriving this equation, we have used the relation

$$\Gamma_0 = \sqrt{G^2(T_0) + |A|^2} = G(T_\infty). \tag{14}$$

While from Eq. (2), we have

$$n_0(T_0) = \frac{f(T_0)}{f(T_\infty)}. \tag{15}$$

Substituting Eq. (8) into Eq. (13) we find:

$$2i\frac{\partial A}{\partial z} + \nabla^2_\perp A + \frac{2}{G(T_\infty)}[1 - n(T_0)]A = 0, \tag{16}$$

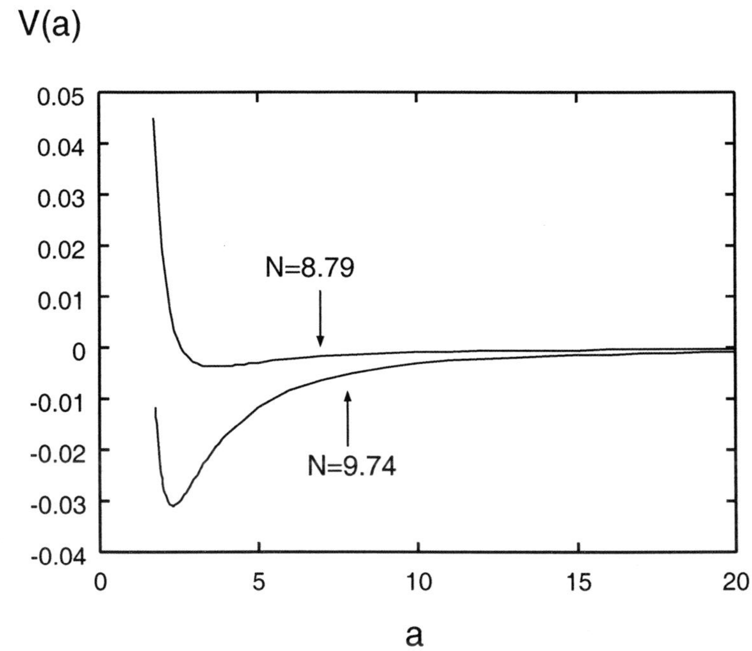

Figure 2: Potential profile of $V(a)$ versus a in the case of $N = 8.79$ and $N = 9.74$.

where subscripts for variables (x_1, y_1, z_2) are dropped for simplicity. We also assumed without loss of generality that $(a^2 - 1)/\epsilon^2 a^2 = 2/G(T_\infty)$, which in dimensional units (provided that $a = \omega/kc$) coincides with the linear dispersion relation of the EM wave in an e-p plasma:

$$\omega^2 = k^2c^2 + \frac{2\omega_e^2}{G(T_\infty)}. \tag{17}$$

Thus, the dynamics of EM beams in hot relativistic e-p plasma has become accessible within the context of a generalized nonlinear Schrödinger equation (16).

We first seek the localized 2D soliton solutions of Eq. (16), and analyze the stability of such solutions. Making the self-evident re-normalization of variables $z \to zG(T_\infty)$, $r_\perp \to r_\perp\sqrt{G(T_\infty)/2}$, Eq. (16) can be written as:

$$i\frac{\partial A}{\partial z} + \nabla_\perp^2 A + \Psi A = 0, \tag{18}$$

where $\Psi = 1 - n_0(T_0)$ represents the generalized nonlinearity. The companion equation (15) can be viewed as a transcendental algebraic relation between T_0 and $|A|^2$, i.e. Ψ is an implicit function of $|A|^2$ $[\Psi = \Psi(|A|^2)]$. We note that Eq. (18) can be written in the Hamiltonian form $i\partial_z A = \delta H/\delta A^*$, where the Hamiltonian is

$$H = \int d\mathbf{r}_\perp[|\nabla_\perp A|^2 - F(|A|^2)], \tag{19}$$

and $F(t) = \int_0^t \Psi(t')dt'$. This implies that Eq. (18) conserves the Hamiltonian H in addition to the power (photon number) $N = \int d\mathbf{r}_\perp |A|^2$.

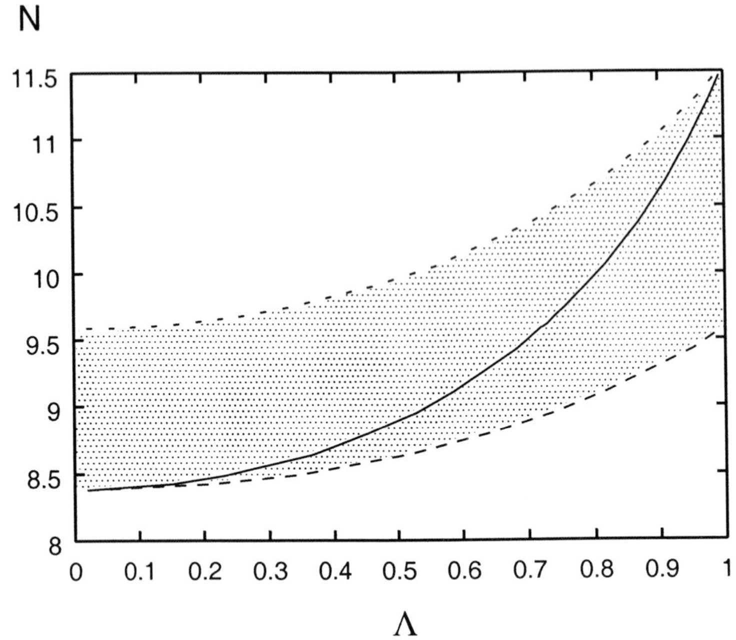

Figure 3: The parameter regime where solitary wave can be trapped for $T_\infty \ll 1$.

To find a stationary, nondiffracting axisymmetric solution, we use the representation of vector potential $A/A_c = U(r)\exp(i\lambda z)$, where $A_c = \left[G(T_\infty)^2 - 1\right]^{1/2}$, $r = (x^2 + y^2)^{1/2}$ and λ is the nonlinear wave-vector shift. The radially dependent envelope $U(r)$ obeys a nonlinear ordinary differential equation:

$$\frac{d^2U}{dr^2} + \frac{1}{r}\frac{dU}{dr} - \lambda U + \Psi(U^2)U = 0. \tag{20}$$

The profiles of the field $U(r)$, the plasma density $n_0(r)$, and the temperature $T_0(r)$ of the stationary solution can be found in Fig. 3 of Ref. [19] for $\lambda = 0.1$; The plasma temperature and density is reduced in the region of field localization. Similar plots could be obtained for all allowed values of λ. When $\lambda \to \lambda_c$, where λ_c is the upper bound of the propagation constant, the plasma cavitation takes place, i.e. the plasma density and temperature tends down to zero at $r = 0$. Appearance of zero temperature is not surprising since the corresponding region is the "plasma vacuum"; all particles are gone away.

The stability of obtained soliton solutions can be investigated using the stability criterion of Vakhitov and Kolokolov [21]. According to this criterion the soliton is stable against small arbitrary perturbations if

$$\frac{dN}{d\lambda} > 0, \tag{21}$$

where N is the photon number or more precisely the power of the trapped mode. In our case, the dependence of N on λ is shown in Fig. 1. One can see that $dN/d\lambda > 0$ everywhere and consequently the corresponding solution is stable for $0 < \lambda < \lambda_c$.

3 Nonlinear Dynamics Based on Variational Approach

The complex dynamics of a beam governed by Eq. (18) can be analyzed by the variational approach [22]. This approach determines the relations between the characteristic parameters of the localized solution approximated by a trial function. The variational method gives qualitatively good results, provided the beam does not undergo structural changes during its evolution. The first standard step is to construct the Lagrangian

$$L = \frac{1}{2}\left(A^* \frac{\partial A}{\partial z} - \text{c.c.}\right) - \mathcal{H}, \tag{22}$$

where $\mathcal{H}$ is the Hamiltonian density [$H = \int d\mathbf{r}_\perp \mathcal{H}$, see Eq. (19)]. In the optimization procedure, the first variation of the variational function must vanish on a suitably chosen trial function. As a trial function, we will use the Gaussian-shaped beam,

$$A = \Lambda(z)\exp\left[-\frac{r^2}{2a(z)^2} + ir^2 b(z) + i\phi(z)\right], \tag{23}$$

with the amplitude Λ, the beam radius a, the wave front curvature b and the phase ϕ as the unknown functions of the propagation coordinate z respectively, which will be furthermore used in order to make the variational functional an extremum. Substituting expression (23) into Eq. (22) and demanding that the variation of the spatially averaged Lagrangian with respect to each of these parameters is zero, we obtain the corresponding set of Euler-Lagrange equations,

$$\frac{d}{dz}\left(\Lambda^2 a^2\right) = 0, \tag{24}$$

$$\frac{d^2 a}{dz^2} = \frac{4}{a^3} - \frac{2}{a}\left[K'(\Lambda^2) - \frac{K(\Lambda^2)}{\Lambda^2}\right], \tag{25}$$

$$b = \frac{1}{4a}\frac{da}{dz}, \tag{26}$$

to be solved for the three functions Λ, a, and b, where the function $K(u)$ is defined as

$$K(u) = 4\int_0^\infty dp\, pF\left(ue^{-p^2}\right). \tag{27}$$

Equation (24) is nothing but a statement of the fact that during the EM beam evolution its power is conserved,

$$N = \pi\Lambda^2 a^2 = \pi\Lambda_0^2 a_0^2, \tag{28}$$

where Λ_0 and a_0 are respectively the initial amplitude and the initial "radius" of the EM beam at $z = 0$.

Using Eq. (24), the integration of Eq. (25) leads to

$$\frac{1}{2}\left(\frac{da}{dz}\right)^2 + V(a) = H = V(a_0), \tag{29}$$

where

$$V(a) = \frac{2}{a^2} - \frac{2a^2}{a_0^2\Lambda_0^2}K\left(\frac{\Lambda_0^2 a_0^2}{a^2}\right) \tag{30}$$

plays the role of an effective potential for the evolution of the radius a. We have assumed that the initial beam have a plane front (or zero curvature) $[da/dz|_{z=0} = 0 = b(0)]$. Using the analogy with a particle in a potential well, we can acquire a deeper physical understanding of light beam dynamics. Choosing the initial beam radius a_0 to be equal to the equilibrium radius a_e, a stationary solution of Eq. (29) is obtained if $\partial V/\partial a|_{a=a_e} = 0$. Note that $-\partial V/\partial a$ is equal to the right-hand side of Eq. (25). The equilibrium radius of the beam is readily found to be

$$a_e^2 = 2\left[K'(\Lambda_0^2) - \frac{K(\Lambda_0^2)}{\Lambda_0^2}\right]^{-1}. \tag{31}$$

In the subsequent analysis we will consider the small temperature case $T_\infty \ll 1$, and the case of moderately high temperature $T_\infty \sim 1$. We do not consider here the ultrarelativistic temperature case ($T_\infty \gg 1$) because applied model equations fail to adequately describe the plasma dynamics due to the neglect of heavy particle production.

In the small temperature case the nonlinear term in Eq. (18) reduces to a simple analytic expression

$$\Psi\left(|A|^2\right) = 1 - \left(1 - |A|^2\right)^{3/2}, \tag{32}$$

which gives the function $F(u)$ as

$$F(u) = u + \frac{2}{5}(1-u)^{5/2} - \frac{2}{5}. \tag{33}$$

We will apply the established general formalism to the saturating nonlinearity given by Eq. (32). The function K becomes [see Eq. (27)]

$$\begin{aligned} K(u) = 2u + \frac{4}{5}\left[(1-u)^{1/2}\left(\frac{46}{15} - \frac{22u}{15} + \frac{2u^2}{5}\right)\right] \\ - \frac{8}{5}\arctan(1-u)^{1/2} - \frac{4}{5}\log u - \frac{4}{5}\left(\frac{46}{15} - \log 4\right). \end{aligned} \tag{34}$$

Note that normalized strength of the field $A\,(= A/A_c)$ is restricted from above $|A| \leq 1$. Above this value the wave-breaking of the field takes place. Substituting Eq. (34) into Eq. (25), we can investigate the nonlinear dynamics of the beam for different initial conditions.

Using Eq. (34) we can find the effective potential $V(a)$ [see Eq. (30)]. The shape of effective potentials in the case of $N = 8.79$ and $N = 9.74$ are illustrated in Fig. 2. From the shape of the potential we conclude that the "effective particle" (i.e. the beam) can be trapped in the potential well. The endpoints of these lines (for small a) correspond to the occurrence of zero density. As a decreases the beam amplitude Λ increases and approaches the unity at endpoints. Note that the potential at the endpoint is negative for $N = 9.74$. Thus, the cavitation can take place when the initial radius a_0 is large enough. It is obvious that $a(z)$ is an oscillating function provided that a_0 is in the trapping region.

The beam never diffracts if $H < 0$ which coincides with the general criterion established by Zakharov *et al.* [23]. Figure 3 shows the parameter regime where the solitary wave could be trapped. The central line denotes the equilibrium curve, and the shaded region between two dashed curves denotes a region where EM beam is trapped. The lower

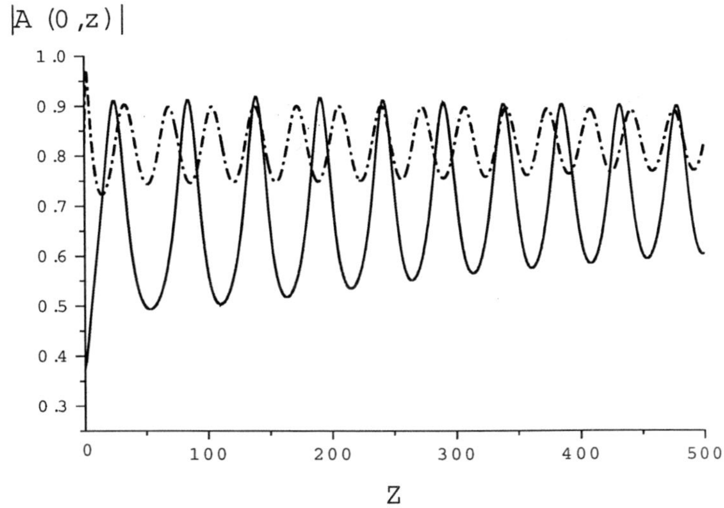

Figure 4: The beams parameters are oscillating around the equilibrium with initial focusing ($\Lambda_0 = 0.37$) or defocusing ($\Lambda_0 = 0.95$).

dashed line corresponds to the zero Hamiltonian. Thus, below this line, the electromagnetic beam with any parameters will diffract. Above the upper dashed line, the electromagnetic beam amplitude will grow up to wave-breaking limit. Therefore, the beam will be trapped in oscillatory regime provided that its parameters are situated in shaded region in Fig. 3.

4 Numerical Simulation

In the preceding analysis, we applied a variational approach involving a Gaussian trial function. The main limitation of this approach is that it is valid only in the aberration-less approximation, i.e. the approach is unable to account for structural changes in the beam shape. Such aspects of the beam dynamics are better delineated by numerical simulations. The detailed dynamics of arbitrary field distribution must be studied by direct simulations of Eq. (18). The guidelines for simulation are still provided by approximative analytical approaches.

The initial profile of the beam is taken to be Gaussian $A = \Lambda_0 \exp(-r^2/2a_0^2)$. In the case of small temperature $T_\infty = 0.01$ the initial parameters are taken in the trapping region shown in Fig. 3. The initial focusing or defocusing before reaching the equilibrium, can be seen in Fig. 4, where the z-propagations of the beam fields $|A(r = 0, z)|$ with same initial power $N = 9.5$ but with different amplitudes $\Lambda_0 = 0.37$ and $\Lambda_0 = 0.95$ are simultaneously drawn. Corresponding initial states of the beams are respectively situated on the left and right side of equilibrium curve in Fig. 3. Figure 4 shows that the beams parameters are oscillating around the equilibrium with initial focusing ($\Lambda_0 = 0.37$) or defocusing ($\Lambda_0 = 0.95$) in agreement with the prediction of variational approach.

Figure 5 shows the field distribution versus radius r and the propagating coordinate z for $\Lambda_0 = 0.37$. Note, however, that due to the appearance of the radiation spectrum the amplitudes of the field oscillations are monotonically decreasing with increasing z. For larger

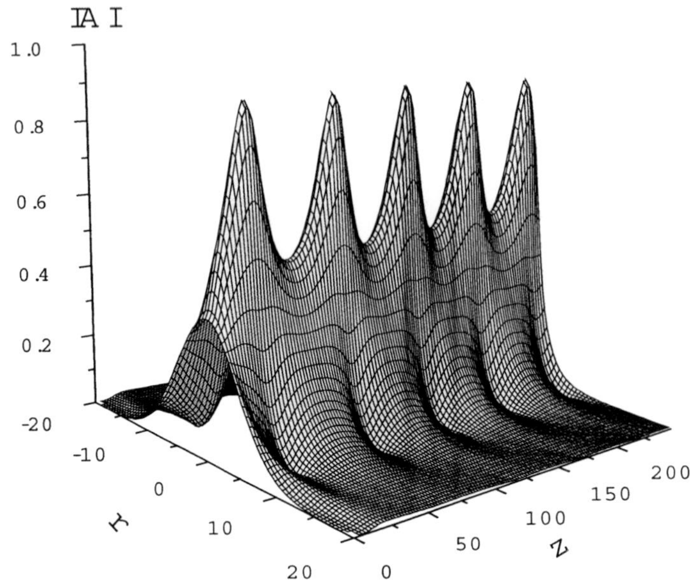

Figure 5: The field distribution versus r and z for $\Lambda_0 = 0.37$.

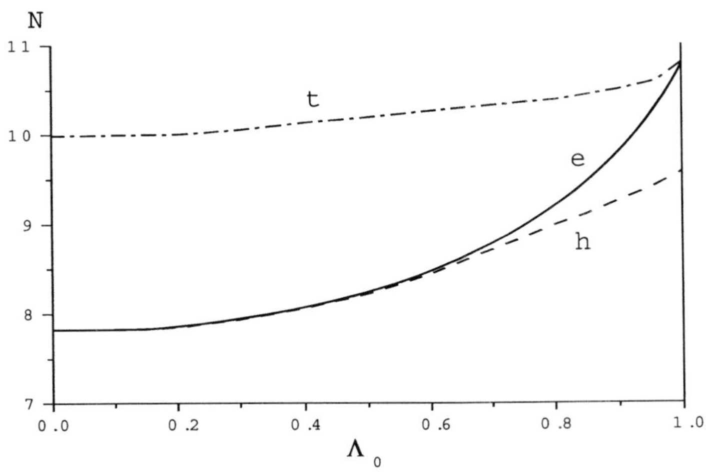

Figure 6: The trapping region in (N, Λ_0) plane obtained by numerical simulations for $T_\infty \ll 1$. The curve 'e' corresponds to the equilibrium state, the curve 'h' the zero Hamiltonian, and the curve 't' the trapping boundary, respectively.

z the formation of a ground solitonic state may take place due to the damp-out of the oscillations. If the initial profile of the beam is close to the equilibrium one, then the beam quickly reaches the profile of ground-state equilibrium, and propagates for a long distance without much distortion of its shape. Since variational approach is unable to account the structural changes in the beam shape and corresponding formation of radiation spectra we can expect that the trapping region in (N, Λ_0) plane obtained by numerical simulations will be differ-

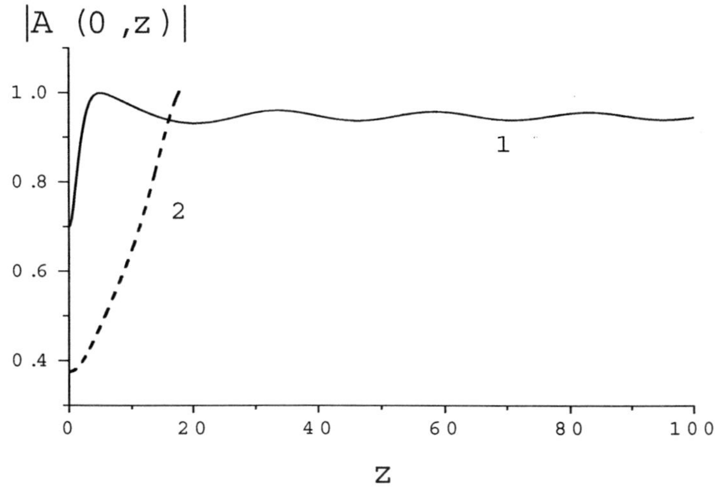

Figure 7: The field $|A(r = 0)|$ versus z for the cases when the beams parameters are in below (the curve 1) and above (the curve 2) the line t.

ent from the one shown in Fig. 3. The result of these simulations is presented in Fig. 6. The curve 'e' corresponds to the equilibrium state, the curve 'h' the zero Hamiltonian, and the curve 't' the trapping boundary, respectively. The Gaussian beam with initial parameters (i.e. N and Λ_0) in the region below line h has a positive Hamiltonian and will be diffracted. The trapping region of the beam is the area between lines h and t. The beam with the parameters in this region will be trapped in self-guiding regime of propagation and will either focus or defocus to the ground state, exhibiting damped oscillations around it. The initial focusing (defocusing) takes place if the parameters are in the region between lines t and e (between e and h). Note that the area of the trapping region shown in Fig. 6 is larger than it follows from variational approach (see Fig. 3). This enlargement of the trapping area is related to the radiation losses during the beam convergence to the equilibrium state.

The beam with parameters in region above trapping line t will focus until wave-breaking and plasma cavitation takes place. Figure 7 shows the plot $|A(r = 0)|$ versus z for the cases when the beams parameters are in the regions below (the curve 1) and above (the curve 2) the line t. One can see that in the first case the beam trapping takes place while, in the second case the beam amplitude increases up to the wave-breaking limit ($A = 1$).

For relativistic high temperature case the EM beam dynamics is similar to what we observed in the low temperature case. Figure 8 shows the trapping region for $T_\infty = 0.3$ (the notations are same as in Fig. 6). Quantitative difference stems from the fact that in the high temperature case the mass of the e-p pairs are modified by the temperature dependent G-factor. It is interesting to mention that in this case in the region of field localization the plasma temperature can be decreased considerably. In Fig. 9 we demonstrate $T(r)$ versus z during the trapped beam evolution toward the stable equilibrium.

It is thus confirmed by numerical simulations that the beam trapping is robust especially for low background temperature regime. The beam can achieve the equilibrium state from a wide range of parameters, even far from the equilibrium, as it is predicted by our analytic approach. However, the area of trapping region reduces as temperature increases.

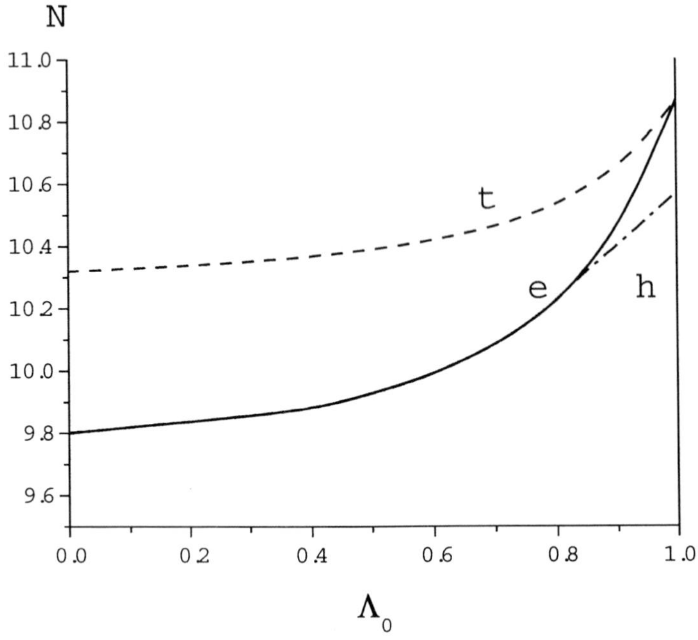

Figure 8: The trapping region in (N, Λ_0) plane for $T_\infty = 0.3$ (the notations are same as in Fig. 6).

5 Summary

We have investigated the nonlinear propagation of strong 2D EM radiation in a relativistic, unmagnetized electron-positron plasma. The treatment is fully relativistic — in the thermal motion as well as in the coherent motion of the plasma particles. The fact that relativistically hot e-p plasmas are capable of sustaining high amplitude localized structures of high amplitude electromagnetic fields should be important to understand the complex radiative properties of different astrophysical objects where such plasmas are considered to exist.

By applying a variational technique with a Gaussian trial function, we have demonstrated the possibility of different regimes of EM beam propagation. In particular, the trapped beam exhibits an oscillatory behavior around the stationary solution. However, since the system contains an upper bound of the field amplitude, the parameter regime is restricted for these oscillations. If the initial beam intensity is large enough, the ponderomotive force acts strongly leading to the beam over-focusing, and eventually the EM field gives rise to the plasma cavitation and wave-breaking.

The region in parameter space where the beam can be regularly trapped and oscillate is found. We have also performed the numerical simulations and demonstrated the parameter regime where beam trapping takes place. Comparing the results of variational approach and the results of numerical simulations we have confirmed that analytical estimate gives a qualitatively good prediction for nonlinear wave dynamics. Due to the radiation and self-deformation of the beam, however, it is shown that the beam may be more robust than it follows from variational method.

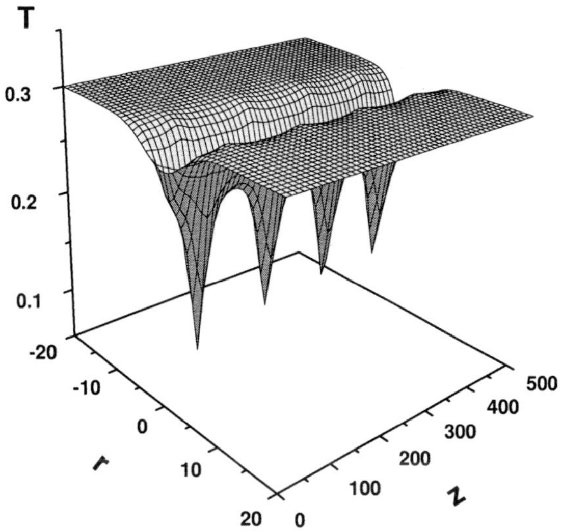

Figure 9: The plasma temperature $T(r)$ versus z during the trapped beam evolution toward the stable equilibrium.

We may apply the present results to the problem of gamma-ray bursts in the pulsar magnetosphere. Lyutikov and Blackman address the fundamental but unresolved issue — how Poynting flux is converted into gamma-rays and the subsequent emission characteristics [18]. Poynting-flux driven outflows from magnetized rotators are a plausible explanation for gamma-ray burst engines. These authors argue that, in a region near the rotation axis, the Poynting flux drives the large-amplitude electromagnetic waves "break" at radii $r_t \sim 10^{14}$ cm, and consequently the MHD approximation becomes invalid. In the "foaming" (i.e., relativistically re-connecting) regions formed during the wave breaking, the random electric fields stochastically accelerate particles to ultra-relativistic energies which then generate turbulent electromagnetic fields. Thus, the claim is that due to wave breaking such a radiation can be considered as a source of gamma-ray bursts. Our results indicate that the wave-breaking can take place at a distance much shorter than r_t.

The work of VIB and SVM has been supported by ISTC Grant G663.

References

[1] P.A. Sturrock, *Astrophys. J.* **164**, 529 (1971); M.A. Ruderman and P.G. Sutherland, *ibid.* **196**, 51 (1995); F.C. Michel, *Theory of Neutron Star Magnetospheres* (University of Chicago Press, Chicago, 1991).

[2] M.C. Begelman, R.D. Blandford, and M.J. Rees, *Rev. Mod. Phys.* **56**, 255 (1984).

[3] M.L. Burns, in *Positron-Electron Pairs in Astrophysics* edited by M.L. Burns, A.K. Harding, and R. Ramaty (American Institute of Physics, New York, 1983).

[4] S. Weinberg, *Gravitation and Cosmology* (Wiley, New York, 1972); P.J.E. Peebles, *Principles of Physical Cosmology* (Princeton University Press, Princeton, 1993).

[5] M. Perry and G. Mourou, *Science* **264**, 917 (1994); S.P. Hatchett *et al.*, *Phys. Plasmas* **7**, 2076 (2000).

[6] A.C.L. Chian and C.F. Kennel, *Astrophys. Space Sci.* **97**, 9 (1983).

[7] R.E. Kates and D.J. Kaup, *J. Plasma Phys.* **41**, 507 (1989).

[8] R.T. Gangadhara, V. Krishan and P.K. Shukla, *Mon. Not. Roy. Astron. Soc.* **262**, 151 (1993).

[9] T. Tajima and T. Taniuti, *Phys. Rev. A* **42**, 3587 (1990); K. Holcomb and T. Tajima, *Phys. Rev. D* **40**, 3909 (1989).

[10] V.I. Berezhiani and S.M. Mahajan, *Phys. Rev. Lett.* **73**, 1110 (1994); V.I. Berezhiani and S.M. Mahajan, *Phys. Rev. E* **52**, 1968 (1995).

[11] T. Tatsuno, V.I. Berezhiani, and S.M. Mahajan, *Phys. Rev. E* **63**, 046403 (2001); T. Tatsuno, V.I. Berezhiani, M. Pekker, and S.M. Mahajan, *Phys. Rev. E* **68**, 016409 (2003).

[12] M.J. Rees, *preprint* (http://xxx.lanl.gov/abs/astro-ph/9701162) (1997).

[13] V.V. Usov, *Nature* **357**, 472 (1992).

[14] C. Thompson, *Mon. Not. Roy. Astron. Soc.* **270**, 480 (1994).

[15] E.G. Blackman, L. Yi, and G.B. Field, *Astrophys. J.* **473**, L79 (1996).

[16] P. Mészáros and M.J. Rees, *Astrophys. J.* **482**, L29 (1997).

[17] W. Kluzniak and M.A. Ruderman, *Astrophys. J. Lett.* **505**, 113 (1998).

[18] M. Lyutikov, E.G. Blackman, *Mon. Not. Roy. Astron. Soc.* **321**, 177 (2001).

[19] V.I. Berezhiani, S.M. Mahajan, Z. Yoshida, and M. Ohhashi, *Phys. Rev. E* **65**, 047402 (2002).

[20] G.Z. Sun, E. Ott, Y.C. Lee, and P. Guzdar, *Phys. Fluids* **30**, 526 (1987).

[21] N.G. Vakhitov and A.A. Kolokolov, *Izv. Vuzov Radio Fiz.* **16**, 1020 (1973) [*Sov. Radiophys.* **9**, 262 (1973)].

[22] V. Skarka, V.I. Berezhiani, and R. Miklaszewski, *Phys. Rev. E* **56**, 1080 (1997).

[23] V.E. Zakharov, V.V. Sobolev, and V.C. Synakh, *Zh. Eksp. Teor. Fiz.* **60**, 136 (1971) [*Sov. Phys. JETP* **33**, 77 (1971)].

In: New Developments in Soliton Research
Editor: L.V. Chen pp. 101-121
ISBN 1-59554-561-8

Chapter 4

DIFFERENTIAL-DIFFERENCE EQUATIONS AND THEIR DEPINNING TRANSITIONS

***P.G. Kevrekidis*[1] *and Y. Zolotaryuk*[2]**
[1]Department of Mathematics and Statistics,
University of Massachusetts, Amherst, MA 01003-4515, USA
[2]Group of Mathematical Physics, IMM, Technical University of Denmark,
Building 321, Richard Petersens Plads,
DK-2800 Kgs. Lyngby, Denmark

Abstract

The reaction-diffusion as well as Klein-Gordon differential-difference equation (DDE) models are revisited and categorized as generic [in the presence of a Peierls-Nabarro (PN) barrier for their static solutions] and non-generic (in the absence of one such). The various available ways to construct models without PN barriers are detailed and their respective features are highlighted. This categorization of DDE's is used to identify the properties of travelling waves that are present in these models upon the action of an external force. The onset of the relevant, so-called depinning transition (to travelling waves) in both the generic and the non-generic discretization case is analyzed and its critical properties are elucidated using analytical calculations (normal form, energy landscape and formal expansion considerations), as well as numerical computations.

1 Introduction

Systems with interplay between discreteness, dispersion, and nonlinear interactions are abundant in physical and biological applications. Examples include spin dynamics in magnetic crystals [1], nonlinear charge and excitation transport in solids [2] and biological systems [3, 4, 5, 6], fluxon propagation in Josephson junctions [7], electromagnetic energy propagation in superlattices [8, 9, 10, 11] and photonic crystals [12, 13]. As a result of this interplay, interesting types of nonlinear excitations may appear, such as kink-like heteroclinic connections, as well as pulse-like homoclinic ones.

One of the most important problems in these discrete systems is whether travelling waves are possible. The above mentioned coherent structures can bottleneck the energy

[14]. However, if they are also able to transport it (especially if so in a targeted fashion [15]), then this would make them natural candidate mechanisms for many bioenergetics processes such as photosynthesis [16].

On the other hand, it is well-known that generic systems of differential-difference equations can *not* sustain propagating solutions due to resonance mechanisms [17, 18] (except, possibly, for selected values of velocities[19, 20, 21]). A recent, rigorous such non-existence result in a special case was given in [22]. The existence of the so-called Peierls-Nabarro (PN) energy barrier [23] and the wobbling of the nonlinear wave in it with a characteristic frequency ($\omega = 2\pi v/h$, where v is the speed of the wave and h the spacing of the lattice) induce the resonance of this frequency with the continuous spectrum of extended small amplitude waves. This results in the generic loss of the localized coherent structure kinetic energy in favor of the small amplitude "radiation". In this way, the waves (and in particular the fronts or kinks that will be the main focus of this study) come to a halt and become pinned between two sites of the lattice. Before continuing, it is instructive to explain the origin of the PN barrier. The latter is the outcome of discreteness breaking the translational invariance of the continuum problem and creating an effective potential with the period of the lattice spacing (due to the integer shift invariance of the lattice). The corresponding energy landscape with the energy minima (stable states) and the energy maxima (unstable states) is the PN potential.

The above discussion is of further interest because one can think of *shearing* the above barrier. If an external force term is added to the equation (or some other "appropriate" –we will see what that means later– perturbation is added) then the PN barrier can be sheared, leading eventually to the disappearance of the static solutions through saddle-node bifurcations [24, 25, 26] and the emergence of a washboard energy landscape that *permits* travelling. It is then natural to examine these systems with the external field (which can emulate e.g., an external electric field in many of the relevant applications), to observe how the so-called "depinning" transition (from static solutions into travelling ones) takes place. The "depinning" transition is an old problem in condensed matter physics with a lot of applications in many systems, like vortices in superconductors[27], domain walls and dislocations [28] to name a few. Numerous studies have been devoted to this transition [24, 25, 26, 29] in discrete systems. In an effort to present a complete picture of the onset of travelling (for both Hamiltonian and dissipative systems), which is the main goal of this paper, we will briefly present the previous findings and adapt them also to Hamiltonian systems (as the majority of the relevant works have addressed the case of reaction-diffusion equations).

On the other hand, there are various cases of *non-generic* discrete systems. Under this classification, we place systems with some form of discrete symmetry (namely momentum-conservation, or equivalently an effective form of translational invariance). One such example are integrable discrete systems (see e.g., [30]). However, in the latter case, the theory of inverse scattering can provide us with exact travelling solutions, hence we will not consider such examples. Instead, we will focus on 3 categories of discrete systems that are momentum conserving and consider the onset of travelling in these cases, demonstrating its differences from the case of generic discretizations. In so doing, we aim at understanding the various potential scenarios for the onset of travelling for discrete fronts in Hamiltonian (in particular nonlinear Klein-Gordon) and dissipative (reaction-diffusion) setups.

Our presentation will be structured as follows. In section II, we consider the case of generic differential-difference equations (DDE's) and the onset of travelling in them. In section III, we examine the 3 examples of non-generic DDE's and detail their features and rationale. In section IV, we study the depinning transition for the non-generic discretizations and the relevant scalings (both analytically and numerically). In section V, we give additional (numerical) examples of the onset of travelling in a number of physically motivated, parametric Klein-Gordon chains and finally, in section VI, we summarize our findings and conclude.

2 Generic DDE's

The case of generic DDE's (once again: the distinction is made on the basis of the presence -generic- or absence -non-generic- of the PN barrier) has been treated in considerable detail for reaction-diffusion (RD) settings [24, 25, 26, 29]. As a case example one can consider the discrete ϕ^4 (also termed the discrete Allen-Cahn equation in the RD literature [31]) equation

$$\{\dot{u}_n, \ddot{u}_n\} = \Delta_2 u_n + 2(u_n - u_n^3), \tag{1}$$

where $\Delta_2 u_n = (u_{n+1} + u_{n-1} - 2u_n)/h^2$ is the discrete Laplacian, the overdot denotes temporal derivative (with one overdot, we are in the RD setting and the equation is dissipative, while for two the system is Hamiltonian). Another classic example often considered in the bibliography is the so-called Frenkel-Kontorova model [32] (the discrete sine-Gordon equation, where $2(u_n - u_n^3)$ is substituted by $-\sin(u_n)$ in Eq. (1)). As has been detailed [24, 25, 26] (see also [33]), there is a PN barrier in these models which is periodic in space (cosinusoidal to leading order), while its height is $\sim \exp(-\pi^2/h)$ (to leading order asymptotically); however, for exceptions to this monotonic decay, see also [34]. Details of the relevant derivation and references to earlier works that obtained the exponential (in the lattice spacing) height can be found in [37]. The simplest way to derive this result is by using the continuum problem static solution [explicitly known as $\tanh(x - \xi)$] parametrized by the position of its center ξ and use it in the discrete expression for the energy (using the so-called Poisson summation formula [38]).

Considering the above model with an external force F causes the shearing of the energy landscape discussed above due to the fact that the potential energy now becomes

$$P = \sum_{n=-\infty}^{\infty} \frac{1}{2}(u_{n+1} - u_n)^2 + \frac{1}{2}(u_n^2 - 1)^2 - F u_n, \tag{2}$$

(whereas before the external force term was absent). We can thus see that when the forcing is turned on, the energetics of the problem will be converted from the figure in the top panel of Fig. 1, to the case of the middle panel. If the forcing exceeds a critical strength that is determined by the height of the barrier, then the situation will result in the disappearance of the two static states (the stable one which is centered between sites and the on-site centered unstable one) in a saddle-node bifurcation and hence the wave will be *forced* to travel in the washboarded energy landscape of the bottom panel of Fig. 1. The (approximate) energy landscapes of the figure have been created by using the ansatz continuum solution

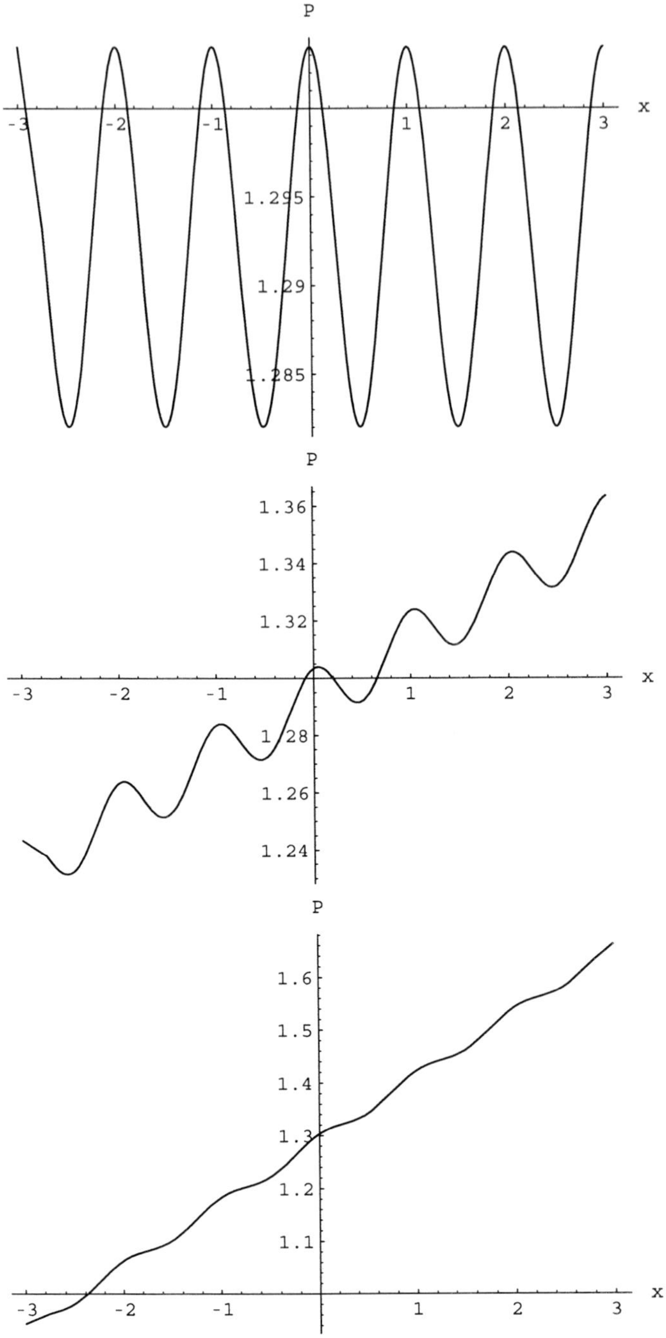

Figure 1: The energy landscape for generic DDE models in the case of $F = 0$ (top panel), $F < F_c$ (middle panel) and $F > F_c$ (bottom panel), as a function of the position of the kink center. Notice that this approximate landscape has been numerically constructed using the static continuum ansatz in Eq. (2) parametrizing the energy by the position of the center of the static front.

mentioned above and looking at the potential energy as a function of ξ. The *exact* potential energy landscape can also be constructed as discussed in [24, 25].

The saddle-node bifurcation was identified in [25] as a saddle-node infinite period (SNIPER) bifurcation in which, at the turning point, there are two distinct heteroclinic connections between the saddle and the node solutions. One of them disappears, whereas the other becomes a homoclinic loop and subsequently a travelling wave (see e.g., [25, 39]). The general structure of the normal forms of a system close to a SNIPER was used in [25] to derive the scaling of the period of the ensuing limit cycle (the time that it takes the front (and each site in particular) to transverse one lattice site) as a function of the distance from the critical point. The relevant scaling is:

$$T \sim (F - F_c)^{-1/2}, \tag{3}$$

from which one can deduce that the speed of the front scales as $C \sim h/T$, hence $C \sim (F - F_c)^{1/2}$, where $1/2$ will be a universal depinning transition exponent for the case of generic DDE's, at the onset of travelling. This was shortly thereafter re-discovered independently in the works of [26]. These authors derived the normal form equation, assuming one "active" site (i.e., that the setting is sufficiently discrete that effectively one site is present in the front's "spine", whereas the others are effectively "at" the steady state value). Adapting their findings to the general setting of both Hamiltonian and RD cases one has:

$$\left\{\frac{du_0}{dt'}, \frac{d^2u_0}{dt'^2}\right\} = \alpha(F - F_c) + \beta u_0^2, \tag{4}$$

where $\alpha = 1 + 1/\{h^2 g'[U_1(F_c)]\} + 1/\{h^2 g'[U_3(F_c)]\}$ and $\beta = -h^2 g''(u_c)/2$. In the above equations t' is an appropriately rescaled time ($t' = t/h^2$ for the RD case, while $t' = t/h$ in the Hamiltonian one), U_1 and U_3 are the steady states of the field for $F = F_c$ and u_c is the value of the active site *at* the critical point (around which we are perturbing). Notice also that the more general form of the model nonlinear equation

$$\{\dot{u}_n, \ddot{u}_n\} = \Delta_2 u_n - g(u_n) + F \tag{5}$$

is assumed. One can easily discern that both in the Hamiltonian and in the dissipative setting the equation Eq. (4) will blow up in finite time. However, it is interesting that that the scalings in the Hamiltonian and dissipative settings are different. In particular, the equation with the first derivative can be rescaled using $u_0 \rightarrow \sqrt{\alpha(F - F_c)/\beta}\mathbf{u_0}$ and $t' \rightarrow \mathbf{t}/\sqrt{\alpha\beta(F - F_c)}$, resulting in $d\mathbf{u_0}/d\mathbf{t} = 1 + U_0^2$ (that has a tangent solution), while for the Hamiltonian setting the scaling of time becomes $t' \rightarrow \mathbf{t}/[\alpha\beta(F - F_c)]^{1/4}$. Though the analytical solution of the latter equation is much more involved, it also blows up (for zero initial conditions) around $t \approx 3.45$. Even though, the derivation of the normal form is no longer valid close to the blow up time, the latter can be used to obtain the scaling of the time of the jump (of the travelling front and of individual sites -see also above-) as a function of the distance from the critical point. The inverse of that time yields the speed of the front with the scaling of

$$C \sim (F - F_c)^{1/2}, \tag{6}$$

$$C \sim (F - F_c)^{1/4}, \tag{7}$$

in the dissipative and Hamiltonian cases respectively. Even though this is in part expected (due to one additional derivative the travelling frame equations have a factor of C and respectively C^2 in the two different settings), we will see that this is not true in the case of non-generic DDE's. It should be noted that the authors of [26] subsequently went on to refine the single active point theory in RD setups, by its multiple active site generalization. However, as can be naturally expected, the scaling properties of the latter were the same as those of the former, resulting in the same scaling of the speed.

3 Non-Generic DDE's

In the previous section, we summarized and extended marginally (to the case of Hamiltonian systems with a shear), the study of the depinning onset for generic DDE's that possess a PN barrier. We now turn our attention to the case of non-generic differential-difference equations for which the PN barrier is absent. The latter will be the main focus of this paper. In particular, we consider the three different discretization construction methods (in the exception of integrable Hamiltonian dynamics) that we are aware of which result in such DDE's:

3.1 Restoring Momentum Conservation

Suppose we consider the model (we restrict ourselves to Hamiltonian systems at present and will return to the consideration of RD dynamics shortly):

$$\ddot{u}_n = \Delta_2 u_n + 2\left(u_n - u_n^2 \frac{u_{n+1} + u_{n-1}}{2}\right). \tag{8}$$

One can suggest that the "inspiration" of this discretization comes from the well-known Ablowitz-Ladik integrable form of the corresponding envelope NLS equation (the so-called AL-NLS model [40]). In fact that is true and this provides also an explicit static solution for this model of the form $u_n = \tanh(an + \xi)$, where $\cosh(2a) = (1 + h^2)/(1 - h^2)$. The model of Eq. (8) does *not* share the integrability of its complex field AL-NLS analog. But it does share an interesting feature that renders the PN barrier *absent* in this discretization. It shares an *effective* form of translational invariance. The latter is mirrored in the presence of a one-parameter set of exact solutions. ξ is free to take any value, hence the restoration of an *effective* translational invariance (TI), albeit in the discrete setting. This invariance should *not* be confused with the integer shift invariance of the equation. Notice that the latter is shared by the model of Eq. (1), but the full restoration of TI present in Eq. (8) is clearly not a feature of Eq. (1). In fact, one can directly connect (through Noether's theory) the effective TI with a momentum conservation law that is present in Eq. (8) but not in Eq. (1). In particular, if we multiply both sides of Eq. (8) by $u_{n+1} - u_{n-1}$, the equation can be re-cast in the following conservation law form:

$$\frac{d\rho_n}{dt} = J(n+1, n) - J(n, n-1), \tag{9}$$

where $d\rho_n/dt = \ddot{u}_n(u_{n+1} - u_{n-1})$ and $J(n + 1, n) = (u_{n+1}^2 + u_n^2)/h^2 - (2 - h^2)u_n u_{n+1}/h^2 - u_n^2 u_{n+1}^2$. Here ρ_n is the momentum density and $J(n+1, n)$ is the corresponding flux and immediately Eq. (9) upon telescopic summation implies conservation of the momentum M

$$M = \sum_{n=-\infty}^{\infty} \dot{u}_n \frac{(u_{n+1} - u_{n-1})}{2h}. \tag{10}$$

The scaling factor of $2h$ has been used to emulate the continuum momentum conservation law which, for these Klein-Gordon equations, assumes the form of $M = \int u_t u_x dx$ being conserved.

One can however generalize the relevant rationale to the discretization of any continuum nonlinearity of order $2k+1$ (i.e., with a nonlinear term $g(u) \sim u^{2k+1}$) in the form of the model

$$\ddot{u}_n = \Delta_2 u_n + u_n^{k+1} \frac{(u_{n+1}^k + u_{n+1}^{k-1} u_{n-1} + \cdots + u_{n+1} u_{n-1}^{k-1} + u_{n-1}^k)}{k+1}. \tag{11}$$

Using the same "integrating factor" $(u_{n+1} - u_{n-1})$, one can readily observe that the equation can be re-cast into the form of Eq. (9) with the same density and the flux:

$$J(n+1,n) = \frac{1}{h^2}(u_{n+1}^2 + u_n^2) - \frac{2}{h^2} u_n u_{n+1} + \frac{1}{k+1} u_{n+1}^k u_n^k. \tag{12}$$

Once again this implies that the momentum conservation law holds in the form of Eq. (10). Notice that using the formulation of Noether's theorem (see e.g., [41]), the one parameter group of diffeomorphisms h^s that are related to the conservation law can be expressed as $h^s(u_n) = u_n + s(u_{n+1} - u_n)/(2h)$. Since the second term in this expression denotes the discretization of u_x, it is clear that the relevant group is related to translations, albeit in the discrete setting. Notice that this one parameter group of diffeomorphisms (and the corresponding one parameter family of solutions) are *not* present in the case of generic DDE's. It should also be clear that the resulting static problem $J(n+1,n) = J(n,n-1) = J(n-1,n-2) = \cdots =$const. has a one-parameter family of solutions since the constant can be computed by the nature of the asymptotic state that is approached [for the case of the front in Eq. (8), $J=0$]. An intuitive way to see that this implies that if a solution exists, it can be centered anywhere between two lattice sites, is to use $J =$const., substitute any value (in the relevant interval of values between the two steady states that the front connects) as u_n. Then from the equation $J(n+1,n) =$const., one can find the value of u_{n+1}. Since this can be done for any u_n [and not only for specific values of u_n as in the model of Eq. (1)], this confirms the effective translational invariance of the discrete model, the existence of a one-parameter family of static solutions [verified in Eq. (8) also by their analytic form], the conservation of linear (discrete) momentum in such models and the resulting absence of a PN barrier (since all positions of the center of the front are equivalent).

The discretization of Eq. (11) was more specifically aimed at odd power nonlinearities (and inspired from the AL-NLS discretization). On the other hand, one can use a discretization scheme that does not distinguish between even and odd nonlinearities, in particular

$$\ddot{u}_n = \Delta_2 u_n + \frac{u_{n+1}^k + u_{n+1}^{k-1} u_{n-1} + \cdots + u_{n+1} u_{n-1}^{k-1} + u_{n-1}^k}{k+1}, \tag{13}$$

in which case it follows that the current density will be

$$J(n+1,n) = \frac{1}{h^2}(u_{n+1}^2 + u_n^2) - \frac{2}{h^2}u_n u_{n+1} + \frac{1}{k+1}(u_{n+1}^{k+1} + u_n^{k+1}). \tag{14}$$

It is worth noting that the methods of this section can be applied not only to power law nonlinearities but to arbitrary ones. For instance, if we consider the equation:

$$\ddot{u} = u_{xx} + f(u), \tag{15}$$

then, if F is the anti-derivative of f, the discretization

$$\ddot{u}_n = \Delta_2 u_n + \frac{F(u_{n+1}) - F(u_{n-1})}{u_{n+1} - u_{n-1}} \tag{16}$$

preserves (under time evolution) the discrete momentum of Eq. (10). The corresponding current density then reads

$$J(n+1,n) = \frac{1}{h^2}(u_{n+1}^2 + u_n^2) - \frac{2}{h^2}u_n u_{n+1} + F(u_{n+1}) + F(u_n). \tag{17}$$

In all of the above cases, the conservation law of Eq. (10) holds, leading to the presence of a neutral eigendirection (due to Noether's theorem) and (equivalently) to a one-parameter family of solutions (an effective TI). Hence, in this class of models (as well as in the two other non-generic classes that will be presented below), the energy will be independent of the position of the center of the front ξ. Once this "flat" energy landscape is "sheared" by an external force, immediately travelling will be initiated (since in this case there is no barrier to "overcome"). Hence, in this case travelling should be present for every value of the shearing parameter F. We will return to this issue in the next section and highlight the differences between Hamiltonian and dissipative models in this respect.

3.2 Restoring a Topological Energy Minimum

Instead of using the above type of discretization (which becomes separable upon the multiplication with the integrating factor) at the equation level, one can use the same argument at the (potential) energy level. In particular, one can use the following argument: the potential energy can be written in the form:

$$P = \sum_n D^2 + F^2, \tag{18}$$

where D is a discretization of the continuum term $u_x^2/2$ (the coupling energy or elastic energy), while $F^2 = V(u)$ is a discretization of the on-site potential, assumed positive definite for the purposes of this method. More specifically, the method presented here will work for odd power nonlinearities of the form $g(u) \sim u^{2k-1}$. In the latter case, one can use the following discretizations:

$$D = \frac{u_{n+1} - u_n}{h} \tag{19}$$

$$F = \frac{u_{n+1}^k + u_{n+1}^{k-1}u_n + \cdots + u_{n+1}u_n^{k-1} + u_n^k}{k+1} \tag{20}$$

The idea behind the latter discretizations is that

$$2DF = \frac{2(u_{n+1}^{k+1} - u_n^{k+1})}{h(k+1)}, \tag{21}$$

and thus given that $P = \sum_n [(D - F)^2 + 2DF]$ (re-writting Eq. (18)), the disappearance of the second term in the bracket, upon the telescopic summation of Eq. (21), leads to a positive-definite potential energy with a lower bound (the so-called Bogomol'nyi bound [42]) for $D = F$. This method was developed and ramified in a series of papers [43], originally for the sine Gordon, then for the ϕ^4 and subsequently for a general Klein Gordon type equation.

Performing the "telescopically motivated" discretization at the energy level also provides (as in the previous subsection) a direct equation for the resulting one-parameter family of solutions (namely $D = F$). Furthermore, this method also creates a discretization without a PN barrier since the energy minimizer equation does not "see" where the nonlinear wave is centered. In particular, if we call $x = nh$, then the minimizer equation satisfies:

$$\frac{\delta E[u(x+\xi)]}{\delta u} = \frac{\delta E[u(x)]}{\delta u} = 0. \tag{22}$$

This statement can be immediately translated upon Taylor expansion into:

$$\left.\frac{\delta^2 E}{\delta u^2}\right|_u \frac{\partial u}{\partial x} = 0, \tag{23}$$

and hence $\partial u/\partial x = \partial u/\partial \xi$ will be an eigenvector of the linearization problem with a corresponding 0 eigenvalue, hence an *effective* TI will be restored in this case as well.

3.3 Discretizing Inversely: Use the Exact Solution, Find the Problem

If a specific solution is sought (such as the hyperbolic tangent solution in models of the type of Eqs. (1) and (8)), one can consider this as an inverse problem and attempt to establish the model that has it as a static solution. In particular, for a given profile $z = u(x)$, one can construct the nonlinearity as

$$g(u) \equiv g(z) = -\frac{1}{h^2}\left[u(u^{-1}(z)+1) + u(u^{-1}(z)-1) - 2z\right]. \tag{24}$$

Then, by construction the solution z satisfies the equation $\Delta_2 u(x) + g[u(x)] = 0, \forall x$. More interestingly perhaps, using $x \to x + \epsilon$, one obtains:

$$\Delta_2 \left.\frac{\partial u}{\partial x}\right|_x \epsilon + g'(u) \left.\frac{\partial u}{\partial x}\right|_x \epsilon = 0. \tag{25}$$

The latter once again implies that $\partial u/\partial x$ is an eigenvector of the linearization with a 0 eigenvalue, and again the model possesses an *effective* TI symmetry and a natural one parameter family of static solutions. This inverse method was used in a number of references [44] but was systematically developed in [20] (see also below).

Another interesting variation, particular to this method is the possibility of obtaining exact travelling solutions in this problem. To achieve that, instead of the above $g(z)$, one has to use

$$g(u) \equiv g(z) = -\frac{1}{h^2}\left[u(u^{-1}(z)+1)+u(u^{-1}(z)-1)-2z\right]+C^2u''(u^{-1}(z)); \quad (26)$$

notice that in this case $x = u^{-1}(z)$ is the travelling frame variable $n - Ct$ and the prime in the above expression refers to differentiation with respect to x. For the example of $u(x) = \tanh(\mu x)$, the corresponding dynamical equation becomes [using Eq. (26)]:

$$\ddot{u}_n = \frac{1}{h^2}\left[u_{n+1}+u_{n-1}+\frac{2(\gamma-1)}{1-\gamma u_n^2}u_n\right]-2\mu^2C^2(u_n-u_n^3), \quad (27)$$

where $\gamma = \tanh^2(\mu)$.

Let us make some observations on this model whose potential was identified in [44, 20] and which was also discussed in [26] for $C = 0$.

- Consider the $C = 0$ case first. The "static" (time-independent) case has been shown to be integrable (as a map) in [45]. Then the hyperbolic tangent is a(n exact) solution and clearly this model should bear a resemblance to the model of Eq. (8). Notice that this is indeed true and the static problems are identical upon multiplication of the former with the factor $1-\gamma u_n^2$. Also, if we multiply both sides by the integrating factor $(1-\gamma u_n^2)(u_{n+1}-u_{n-1})$, we obtain the form of Eq. (9) with

$$J(n+1,n) = [u_{n+1}^2+u_n^2-\gamma u_n^2u_{n+1}^2+2(\gamma-1)u_nu_{n+1}]/h^2 \quad (28)$$

and the conserved quantity analog to the momentum is then

$$M = \int_0^t ds \sum_n [1-\gamma u_n^2(s)]\frac{d^2u_n}{ds^2}[u_{n+1}(s)-u_{n-1}(s)]. \quad (29)$$

- It then becomes intuitively clear why the model with $C \neq 0$ has a PN barrier for its static solutions, as was observed in [20]: the C-dependent perturbation introduces a discretization of the form of Eq. (1) and the PN barrier immediately follows for the static solutions. An alternative, quantitative way to observe that is to consider the travelling solutions (for the case of $C \neq 0$) in the total expression for the energy of the model of Eq. (27):

$$E = \sum_{n=-\infty}^{\infty}\left[\frac{1}{2}\dot{u}_n^2+\frac{1}{2h^2}(u_{n+1}-u_n)^2+V(u_n)\right], \quad (30)$$

(where $V(u_n)$ is explicitly given in Eq. (27) of [20]). If one uses the exact travelling solution (at time $t = 0$, which noticeably coincides with the static solution for that time) in the *full* energy of Eq. (30), then one obtains the straight line center-independent energy plot in Fig. 2. However, what is the difference between the energy of the static solution and the energy of the travelling solution for the particular value $C \neq 0$ of the speed ? Given the similarity of the profile at $t = 0$, the only

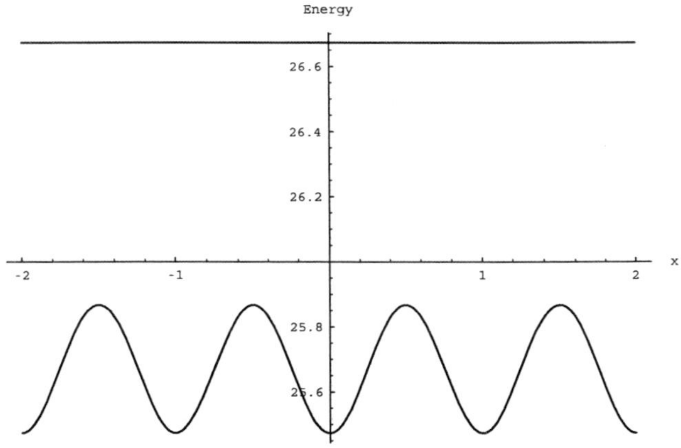

Figure 2: The energy of a travelling solution (horizontal line) and of a static solution (periodic line) as a function of the position of the center of the front calculated through Eq. (30).

difference comes from the kinetic energy contribution in Eq. (30). It is exactly that *kinetic energy contribution* to the full energy that renders the travelling wave solution one such, and which attributes, on the other hand, to the static solution a PN barrier, as is shown in the second graph of the Fig. 2 (showing the -periodic in ξ- energy of the static solution). In fact, using this argument one can explicitly calculate the PN barrier for the static solution since the kinetic energy will be

$$E_k = \sum_{n=-\infty}^{\infty} \frac{1}{2}\dot{u}_n^2 = \frac{C^2\mu^2}{2}\operatorname{sech}^2(\mu(x-\xi)) = \sum_{l=1}^{\infty} \frac{2\pi^2 C^2 l}{\sinh^2(\frac{\pi l}{\mu})}\cos(2\pi l x), \tag{31}$$

and the static energy will be E =const.$-E_k$. It is then clear how the presence of the "speed parameter" C creates a PN barrier that is exponential in the width of the kink μ^{-1} (related to the spacing of the discrete lattice as is customary for PN barriers [23, 24, 25, 26, 29, 33]) but is also quadratic in this speed parameter of the model.

- It is also worth noting here that to create travelling in the static version of the model (satisfying Eq. (24)), it is not necessary to add the speed dependent perturbation of Eq. (26). A simple shearing of the "flat" energy landscape will be enough to create the potential for travelling.

4 The Onset of Travelling for Non-generic Discretizations

In this section, we examine numerically and analytically the onset of travelling for non-generic discretizations, such as the ones presented above where the PN barrier is absent.

Let us start by considering the rather special case of creating *exact* travelling solutions as in the framework of Eq. (26) of the third category of models in the previous section. To examine the propagation of these exact solutions, we simulated the dynamical version of the model of Eq. (26). In Fig. 3, a solution with speed $C = 0.8$ is shown for a time evolution of $t \approx 22.5$. One can observe that at the end of this time interval the solution

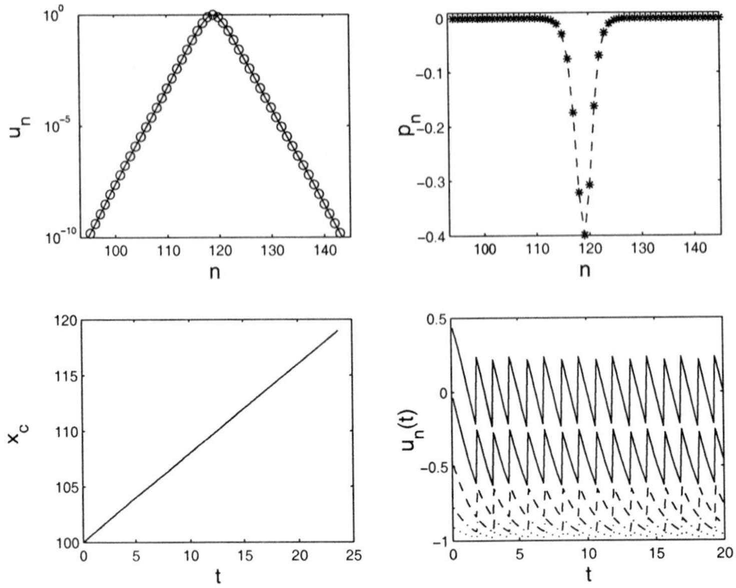

Figure 3: The top left panel shows the solution's spatial profile (actually $|1-u_n^2|$) after $t \approx 22.5$ in a semilog scale. It is clear that it retains its initial shape. The top right panel shows the corresponding momentum distribution for that time, the bottom left panel shows the trajectory of the kink center x_c as a function of time, while the bottom right shows the time evolution of the few sites (nearest ones by solid, next nearest by dashed, then dash-dotted and finally dotted) near the center of the kink. $C = 0.8$.

clearly retains the exponential localization of its spatial profile (top left panel), its original momentum distribution profile (apart from the travelling-induced shift; see top right panel), and the motion of its center has been with uniform speed throughout the interval.

To examine the robustness of travelling in this setting, we considered the following variation in the above scenario. We used a solution with initial speed of 0.8 in the initial conditions, but in the dynamical equation of Eq. (26), $C = 0.5$ was used. This resulted in Fig. 4. Clearly, in this case a lot of strong radiation wakes are shed and the center of the kink is wobbling (something also evident from the motion of the lattice sites near it) in the PN barrier with a speed ≈ 0.65 (as can be deduced from the bottom left panel). In that sense travelling in such models is less robust than in the models with shearing, because in the latter case travelling is initiated for any value of the external force, whereas in the former, slight deviations from the exact conditions that lead to the absence of the PN barrier, regenerate the barrier.

Now let us turn to the generic case of shearing one of the models of the three types presented in the previous section. In the case of RD models, the relevant equation for the ensuing travelling solution will read:

$$-Cu'(x) = \Delta_2 u(x) + g[u(x)] + F \tag{32}$$

But, then using the formal expansion $C = C_0\epsilon$, $F = F_0\epsilon$ and $u(x) = u_n + \epsilon w(x)$

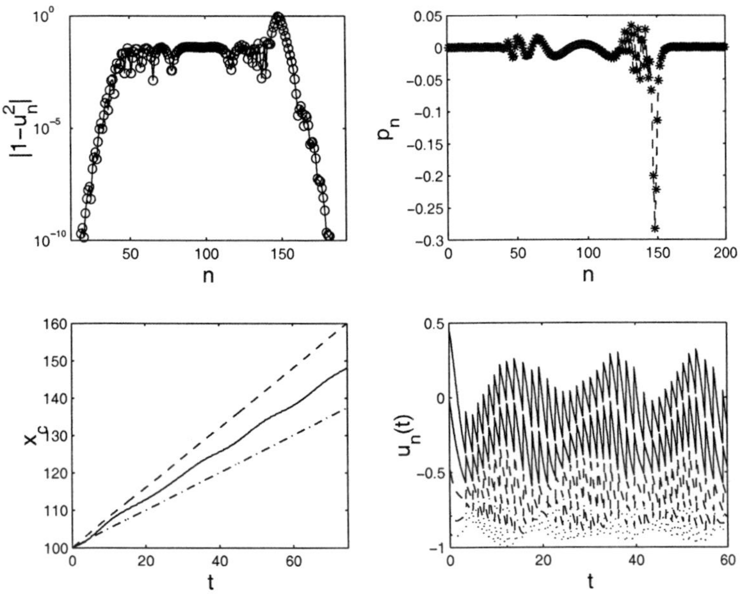

Figure 4: Same as in the previous figure, but for $C = 0.8$ in the initial conditions and $C = 0.5$ in the dynamical equation. Notice the wobbling of the kink center and of the few sites near it in the time evolution of the bottom left and right panels respectively.

(where u_n is the exact static solution for $F = 0$), one obtains:

$$\Delta_2 w(x) + g'(u_n)w(x) = -Cu'_n - F, \tag{33}$$

where $u'_n = u'(x = n)$, and subsequent use of the Fredholm alternative in Eq. (33) yields directly the linear scaling for the front speed as a function of F as:

$$C = -F\frac{U_3 - U_1}{< u'_n|u'_n >}, \tag{34}$$

where $U_{1,3}$ are the left and right end states and the brackets denote inner product. Hence, it is clear that the travelling solutions bifurcate from the static solutions with a linear scaling in the external field, once the RD model is sheared.

However, it is interesting to contrast this case with the Hamiltonian one, where:

$$C^2 u''(x) = \Delta_2 u(x) + g[u(x)] + F. \tag{35}$$

In the latter case, attempting the same trick one obtains that $F(U_3 - U_1) = 0$ (due to parity reasons the speed dependent term vanishes), resulting in the conclusion that $F = 0$ identically. Hence, we can conclude that contrary to the RD case, once sheared the Hamiltonian travelling waves bifurcate from a solution with *nonzero* speed even at $F = 0$. One can then use an expansion of the form: $C = C_0 + \epsilon C_1$, $u(x) = u_0 + \epsilon u_1(x) + \epsilon^2 u_2(x) + \ldots$ One then obtains to leading order:

$$\mathcal{L}u_1 \equiv C_0^2 u_{1,xx} - \Delta_2 u_1 - g'(u_0)u_1 = -2C_0C_1u_{0,xx}, \tag{36}$$

which can be solved for u_1 (using variation of coefficients and the known solutions of the homogeneous problem). $\mathcal{L}$ stands for the linearization operator. A similar equation for u_2 in the next order reads:

$$\mathcal{L}u_2 = -C_1^2 u_{0,xx} - 2C_0 C_1 u_{1,xx} + \frac{1}{2} g''(u_0) u_1^2 + F. \tag{37}$$

Application of the Fredholm alternative in Eq. (37) yields the scaling for F. We won't write down the full resulting equation as it is sufficient to note that $u_1 \sim C_1$ (from Eq. (36)) and hence all the terms (but the last) in Eq. (37) scale as $\sim C_1^2$; thus the external force should also scale according to $F \sim C_1^2$ and hence

$$C \sim C_0 + \delta F^{1/2}, \tag{38}$$

where δ is appropriate constant. Notice that in comparison with the scalings of the generic case of section II, the power of $1/2$ is expected in comparison to the linear scaling of the parabolic case, but the essential difference lies in the presence of a travelling solution *even* at $F = 0$.

To test the predictions of the above theoretical analysis, we considered the model of Eq. (27) but with an external force term F instead of the C-dependent term ($C = 0$ in Eq. (27) and a term "$+F$" is added to the right hand side of the equation). In the numerical simulations the position of the center of the kink was approximated using a center of mass approach for the 10 nearest points to the exact center (for which $u = 0$) of the front. An example of the resulting travelling motion of the center of mass is shown in Fig. 5. The small "wiggles" are a result of the numerical approximation, while the straight line shows the best fit linear approximation that was used to derive the speed of the front. This method was used for deriving the speed for a range of F's between 0 and 0.025. The resulting speeds are shown as a function of F in Fig. 6 (the data points are shown by circles). The solid line indicates the best fit of the form $C = 0.521104 + 0.60378 \times F^{0.21658}$ (with three parameters), while the two-parameter $F^{1/2}$ best fit is shown by dashed line and reads $C = 0.649531 + 0.925439 \times F^{1/2}$. One straighforward observation from both fits is the verification of the theoretical prediction for the existence of a travelling solution *even at* $F = 0$. On the other hand, the exact form of the power law dependence on F is somewhat controversial and clearly merits further numerical investigation. One can observe that the best fit $F^{1/2}$ scaling is not far from the actual result, and the nature of the approximations used in finding the front center position may well justify the discrepancies. On the other hand, the numerical findings seem to be slightly better approximated by a scaling of $F^{1/4}$. Further detailed numerical studies would be useful in clarifying the exact form of the relevant scaling.

5 More Examples: The Case of Monoparametric Nonlinear Klein-Gordon Chains

We now proceed (motivated also by the discussion of the previous section) to the numerical investigation of further examples of kink depinning, in particular in discrete Klein-Gordon

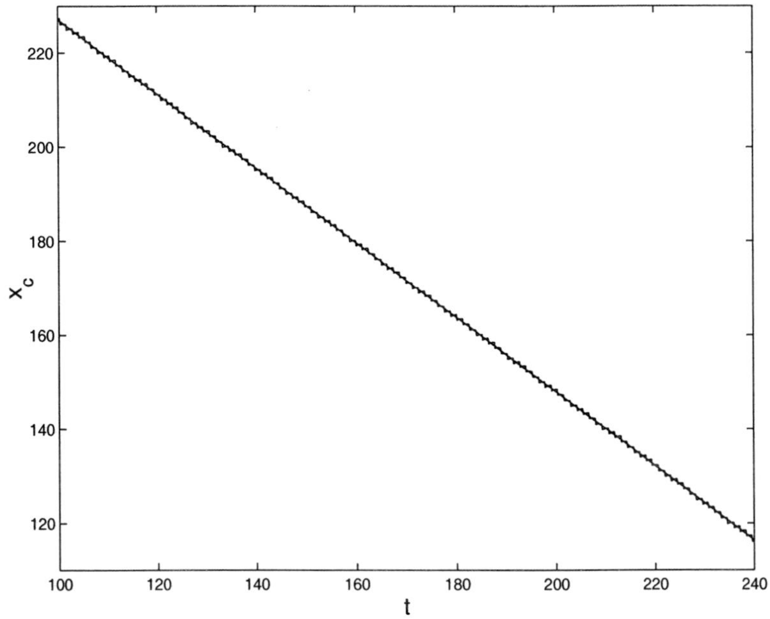

Figure 5: The time evolution of the position of the center of the kink. The straight line (which can practically not be distinguished from the numerical data) is the linear best fit used to extract the speed of the front.

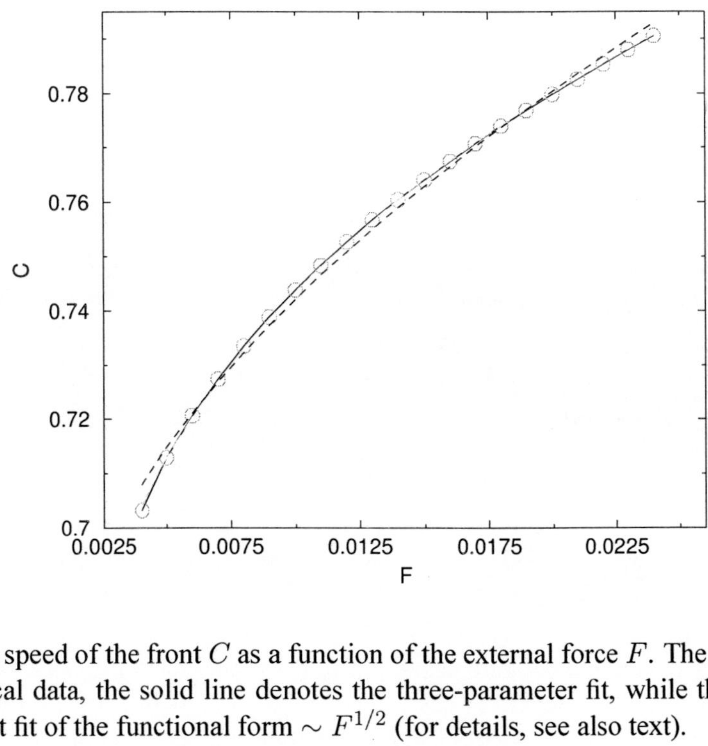

Figure 6: The speed of the front C as a function of the external force F. The circles are the exact numerical data, the solid line denotes the three-parameter fit, while the dashed line shows the best fit of the functional form $\sim F^{1/2}$ (for details, see also text).

models with two physically motivated on-site potentials. Firstly, we consider the double-Morse potential (also known as a sinh-Gordon potential),

$$V(u) = \left[\frac{\cosh\beta - \cosh(\beta u)}{\cosh\beta - 1}\right]^2 , \tag{39}$$

which is used as a first approximation to the proton-ion interaction in the hydrogen-bonded chains (see e.g., [46]). Another potential, introduced by Peyrard and Remoissenet in [35] as a generalization of the sine-Gordon potential:

$$V(u) = \frac{(1+2\beta)(1-\cos u)}{1+\beta(1-\cos u)}, \tag{40}$$

was also examined. In the limit $\beta \to 0$ these potentials tend to ϕ^4 and sine-Gordon potentials, respectively. If β is large enough these potentials have a flat barrier and rather narrow wells and it was shown numerically in Refs. [21, 47] that they allow for travelling waves with some selected velocities. In that sense they belong to the same class of models derived from Eq. (35) with $C \neq 0$. with the only difference that here a travelling kink solution and its velocity are known explicitly. In order to improve the precision of our calculations (and also to avoid undesirable effects of small amplitude phonon radiation) we have added small damping to our equations. Thus, we integrate numerically the damped and dc-driven Klein-Gordon equation

$$\ddot{u}_n - \frac{1}{h^2}(u_{n+1} - 2u_n + u_{n-1}) + \gamma\dot{u}_n + V'(u_n) = F. \tag{41}$$

We compute numerically the kink speed as a function of the external force F. Due to the presence of damping the dynamics of the system will be attracted to the moving kink solutions. In Fig. (7) consider the case of the double-Morse potential, given by Eq. (39). If $\gamma = F = 0$ there are two travelling kink solutions with velocities $C_0^{(1)} = 1.606335$ and $C_0^{(2)} = 0.84168$ at $h = 1/\sqrt{10} \simeq 0.3162$, $\beta = 7$, [47]. Therefore there one should expect two corresponding moving kink solutions in the case $F \neq 0$. The panels (a) and (b) show how the kink velocity depends on the apllied field F for the kink solutions which correspond to velocities $C_0^{(1)}$ and $C_0^{(2)}$, respectively. With decreasing of F both dependences stop at some critical values of $F = F_c$ attaining values close to those $C_0^{(1,2)}$. Values of F_c are rather small and are caused by the presence of dissipation. Obviously, the reason for this is that at some point the force F is not strong enough to overcome the damping. It is expected that F_c will tend to zero as $\gamma \to 0$. The insets show the log-log plots of these dependences. Their fitting yields $C = C_0 + (F - F_c)^\alpha$ with $\alpha = 0.591$ in the case (a) and $\alpha = 0.551$ in the case (b).

In the next figure we show the same dependence, but for the Peyrard-Remoissenet model. For these particular parameters (see caption of Fig. 8) in the limit $\gamma = F = 0$ there exists a travelling kink solution with the velocity $C_0 = 0.590702$ at $h = 1$ and $\beta = 1$. The situation shown in this figure is similar to that in the case of the double-Morse model. Here we considered two values of damping $\gamma = 0.02$ (shown by $\circ$) and $\gamma = 0.005$ (shown by $+$). Fitting the curves gave the following critical exponents: $\alpha = 0.569$ for $\gamma = 0.02$ and $\alpha = 0.563$ for $\gamma = 0.005$. As one can see, numerical computation of the depinning

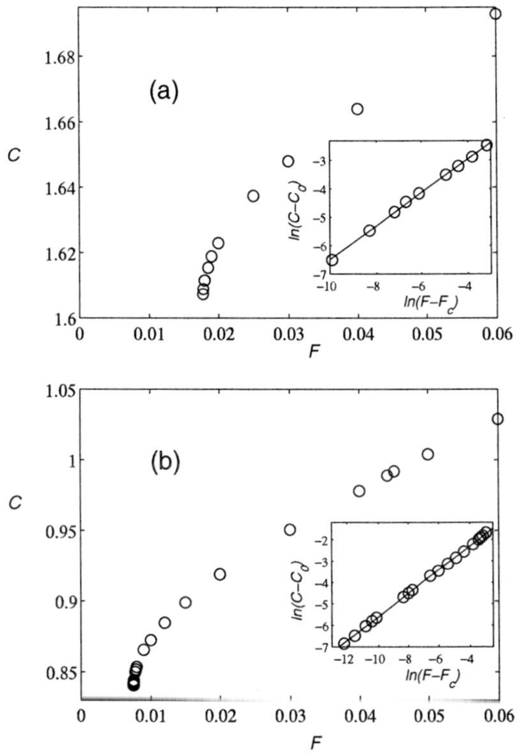

Figure 7: Dependence of the kink velocity on the external force F for the case of double-Morse potenial. Panels (a) and (b) stand for different kink modes (see text). The insets show the corresponding log − log dependences, with solid lines being the best fits (see text for details). Damping in this case was $\gamma = 0.02$.

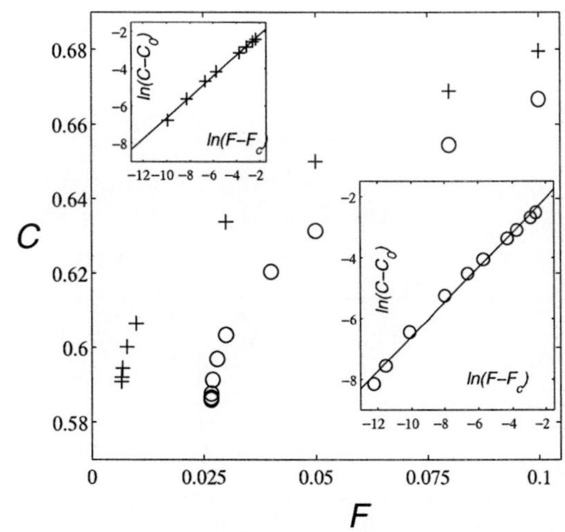

Figure 8: Dependence of the kink velocity on the external force F for the Peyrard-Remoissenet potential for different values of damping: $\gamma = 0.02$ ($\circ$) and $\gamma = 0.005$ ($+$). The insets show the corresponding log − log dependences, with solid lines being the best fits (see text for details).

transition shows that the critical exponent α is always close to the theoretically predicted value $\alpha = 1/2$. These results strongly support the conclusion that depinning from a travelling wave solution at $F = 0$ with this critical exponent is a generic effect, typical for this class of discrete Klein-Gordon models.

6 Conclusions

In this work we have revisited the topic of different discretizations of continuum models and their properties. We have identified the existence of generic discretizations based on the existence of the feature of the Peierls-Nabarro barrier for the corresponding static solutions. We have separated the study of these generic discretizations from the one of non-generic discrete systems "manufactured" in such a way that the Peierls-Nabarro barrier is absent. We have indicated 3 different systematic ways that we are aware of (in the exception of solvability conditions for scattering problems that generate integrable systems), in which such systems may be generated. The first involved a discretization that explicitly restores momentum conservation, the second a discretization that created a lower bound for the discrete system energy, while the third used an inverse method for which the solution was initially determined and then the model was identified that possessed such a solution. In all the cases, the relevant methods restored an *effective translational invariance* in the discrete model, caused the stable and unstable manifolds of the relevant orbits to intersect non-transversely, generated one-parameter families of (static) solutions, possessed a momentum conservation law and the corresponding linearization had an eigenvalue (or a pair of eigenvalues depending on whether the model is dissipative or Hamiltonian) at 0. All of the above properties are inter-connected through Noether's theory, the conservation law formulation and the linearization problem in ways that we have highlighted. Our main goal in separating the generic and non-generic discretizations was to identify their different respective properties in the vicinity of the depinning transition. The criticality and universality of the relevant transition have been explored, as a continuation and expansion of previous works, and the critical exponents have been identified. It has thus been found that for reaction-diffusion systems in agreement with previous findings of [25, 26], a scaling of the speeds $C \sim (F - F_c)^{1/2}$ occurs, while it was identified that in the Hamiltonian case, the relevant scaling is $C \sim (F - F_c)^{1/4}$. This is the universal form of the scaling law for generic discretizations. On the other hand, in the case of non-generic discrete models without PN barrier, the scaling is *linear* in F for reaction-diffusion systems and the relevant solutions bifurcate from the static solutions. On the contrary, it was found that in Hamiltonian systems the relevant bifurcation occurs not from the static solution but from a travelling solution for $F = 0$. This was confirmed through numerical experiments. The relevant scaling in this case was theoretically identified as $C \sim C_0 + \delta F^{1/2}$, consistently with numerical findings. It should be noted in passing that this picture appears consistent with the recent numerical findings of [48], where it was numerically illustrated that travelling solutions in the $F \to 0$ limit do not bifurcate from $C_0 = 0$ (static) limit, despite the absence of the Peierls-Nabarro barrier. Depinning for a broader class of physically motivated models which posess travelling wave solutions has also being investigated numerically and the respective scaling is shown to be around $\delta F^{0.55 \div 0.59}$ which is close to theoretically predicted $1/2$. Detailed numerical examination of such predictions in other systems of the general classes considered herein would be an interesting topic for future studies.

References

[1] R. Lai and A.J. Sievers, *Phys. Rep.* **314**, 148 (1999).

[2] A.S. Davydov, *Solitons in Molecular Systems* (Reidel, Dordrecht, 1985).

[3] A. C. Scott, *Phys. Rep.* **217**, 1 (1996).

[4] M. Peyrard and A. R. Bishop, *Phys.Rev. Lett.* **62**, 2755 (1989); T. Dauxois, M. Peyrard, and A. R. Bishop, *Phys. Rev. E* **47**, 684 (1993).

[5] L.V. Yakushevich, *Nonlinear Physics of DNA* (Wiley, New York, 1998).

[6] Yu. B. Gaididei, S. F. Mingaleev, P.L. Christiansen, M. Johansson, and K. Ø. Rasmussen in *Nonlinear Cooperative Phenomena in Biological Systems* edited by L. Mattson (World Scientific, Singapore, 1998), p. 176.

[7] A. V. Ustinov, *Physica* **123D**, 315 (1998).

[8] *Photonic Band Gaps and Localization*, edited by C.M. Soukoulis (Plenum, New york, 1993); *Photonic Band Gap Materials*, edited by C.M. Soukoulis (Kluwer, London, 1996).

[9] D. Hennig and G. P. Tsironis, *Phys. Rep.* **307**, 334 (1999).

[10] Yu. B. Gaididei, P.L. Christiansen, K. Ø. Rasmussen, and M. Johansson, *Phys. Rev. B* **55**, R13365 (1997).

[11] C. B. Clausen, P. L. Christiansen, L. Torner, and Yu. B. Gaididei, *Phys. Rev. E* **60**, R5064 (1999).

[12] A. R. McGurn, *Phys. Lett. A* **251**,322 (1999); *ibid.* **260**, 314 (1999).

[13] S. F. Mingaleev, Yu. S. Kivshar, and R. A. Sammut, *Phys. Rev. E* **62**,5777 (2000).

[14] T. Cretegny, T. Dauxois, S. Ruffo and A. Torcini, *Physica* **121D**, 109 (1998).

[15] H.E. Nistazakis, P.G. Kevrekidis, B.A. Malomed, D.J. Frantzeskakis and A.R. Bishop, *Phys. Rev. E* **66**, 015601(R) (2002).

[16] G. Kopidakis, S. Aubry, and G.P. Tsironis, *Phys. Rev. Lett.* **87**, 165501 (2001).

[17] M. Peyrard and M.D. Kruskal, *Physica* **14D**, 88 (1984).

[18] P.G. Kevrekidis and M.I. Weinstein, *Physica* **142D**, 113 (2000).

[19] Y. Y. Earmme and J. H. Weiner, *Phys. Rev. Lett.* **33**, 1550 (1974); Y. Y. Earmme and J. H. Weiner, *J. Appl. Phys.* **48**, 3317 (1977).

[20] S. Flach, Y. Zolotaryuk and K. Kladko, *Phys. Rev. E* **59**, 6105 (1999).

[21] A.V. Savin, Y. Zolotaryuk and J.C. Eilbeck, *Physica* **138D**, 267 (2000).

[22] D.E. Pelinovsky and V.M. Rothos, *Bifurcations of travelling breathers in the discrete NLS equations*, preprint (2004).

[23] F.R.N. Nabarro, *Theory of Crystal Dislocations* (Oxford University Press, Oxford, 1967).

[24] I. Mitkov, K. Kladko and J.E. Pearson, *Phys. Rev. Lett.* **81**, 5453 (1998); K. Kladko, I. Mitkov and A.R. Bishop, *Phys. Rev. Lett.* **84**, 4505 (2000).

[25] P.G. Kevrekidis, I.G. Kevrekidis and A.R. Bishop, *Phys. Lett. A* **279**, 361 (2000).

[26] A. Carpio and L.L. Bonilla, *Phys. Rev. Lett.* **86**, 6034 (2001); A. Carpio and L.L. Bonilla (preprint).

[27] G. Blatter *et al*, *Rev. Mod. Phys.* **66**, 1125, (1994).

[28] D. S. Fisher, *Phys. Rep.* **301**, 113 (1998); M. Kardar, *Phys. Rep.* **301**, 85 (1998).

[29] J.P. Keener, *SIAM J. Appl. Math.* **47**, 556 (1987).

[30] M.J. Ablowitz and H. Segur, *Solitons and the Inverse Scattering Transform* (SIAM, Philadelphia, 1981).

[31] D. Bettinson and G. Rowlands, *Phys. Rev. E* **57**, 169 (1998).

[32] J. Frenkel and T. Kontorova, *J. Phys. USSR* **13**, 1 (1938).

[33] P.G. Kevrekidis, C.K.R.T. Jones and T. Kapitula, *Phys. Lett. A* **269**, 120 (2000).

[34] Monotonic decay of the PN barrier as a function of $1/h$ is present in the discrete sine-Gordon and ϕ^4 models. However, for models with rather flat barrier, such as, for example, the modified sine-Gordon (Peyrard-Remoissenet model) [35] or the double-Morse (sinh-Gordon) potentials, this decay is non-monotonic. See also Ref. [36], where an analytical computation of the PN barrier for the PR model is performed.

[35] M. Peyrard and M. Remoissenet, *Phys. Rev. B* **26**, 2886 (1982).

[36] D. Yemélé and T. C. Kofané, *Phys. Rev. E*, **66**, 016606 (2002).

[37] P.G. Kevrekidis and I.G. Kevrekidis, *Phys. Rev. E* **64**, 056624 (2001).

[38] P.M. Morse and H. Feshbach, *Methods of Theoretical Physics* (McGraw-Hill, New York, 1953).

[39] S.H. Strogatz, *Nonlinear Dynamics and Chaos with Applications in Physics, Biology, Chemistry and Engineering* (Addison-Wesley, Reading, MA 1986).

[40] M.J. Ablowitz and J.F. Ladik, *J. Math. Phys.* **16**, 598 (1975); *J. Math. Phys.* **17**, 1011 (1976).

[41] V.I. Arnold, *Mathematical Methods in Classical Mechanics* (Springer-Verlag, New York 1989).

[42] E.B. Bogomol'nyi, *Sov. J. Nucl. Phys.* **24**, 449 (1976).

[43] J.M. Speight and R.S. Ward, *Nonlinearity* **7**, 475 (1994); J.M. Speight, *Nonlinearity* **10**, 1615 (1997); *ibid.* **12**, 1373 (1999).

[44] V.H. Schmidt, *Phys. Rev. B* **20**, 4397 (1979); P.C. Bresslof and G. Rowlands, *Physica* **106D**, 225 (1997).

[45] M. H. Jensen, P. Bak and A. Popielewicz, *J. Phys. A: Math. Gen.* **16**, 4369 (1983)

[46] X. Duan and S. Scheiner, *J. Mol. Struct.* **270**, 236 (1992).

[47] V. M. Karpan, Y. Zolotaryuk, P. L. Christiansen and A.V. Zolotaryuk, *Phys. Rev. E* **66**, 066603 (2002).

[48] A.A. Aigner, A.R. Champneys and V.M. Rothos, *Physica D* **186**, 148 (2003).

In: New Developments in Soliton Research
Editor: L.V. Chen pp. 123-147
ISBN 1-59554-561-8

Chapter 5

GENERALIZED SINE-GORDON AND MASSIVE THIRRING MODELS

Harold Blas
Departamento de Matemática. Univerdidade Federal de Mato Grosso (UFMT) Av. Fernando Correa, s/n, Coxipó, Cuiabá 78060-900, Mato Grosso-Brazil

Abstract

We consider the Lagrangian description of the soliton sector of the so-called affine $\hat{sl}(3)$ Toda model coupled to matter (Dirac) fields (ATM). The theory is treated as a constrained system in the contexts of the Faddeev-Jackiw, the symplectic, as well as the master Lagrangian approaches. We exhibit the master Lagrangian nature of the model from which generalizations of the sine-Gordon (GSG) or the massive Thirring (GMT) models are derivable. The GMT model describes $N_f = 3$ [number of positive roots of $su(3)$] massive Dirac fermion species with current-current interactions amongst all the $U(1)$ species currents; on the other hand, the GSG theory corresponds to $N_b = 2$ [rank of the $su(3)$ Lie algebra] independent Toda fields (bosons) with a potential given by the sum of three SG cosine terms. The dual description of the model is further emphasized by providing some on shell relationships between bilinears of the GMT spinors and the relevant expressions of the GSG fields. In this way, in the **first part** of the chapter, we exhibit the strong/weak coupling phases and the (generalized) soliton/particle correspondences of the model at the classical level. In the **second part** of the chapter we give a full Lie algebraic formulation of the duality at the level of the equations of motion written in matrix form. The effective off-critical $\hat{sl}(3)$ ATM action is written in terms of the Wess-Zumino-Novikov-Witten (WZNW) action plus some kinetic terms for the spinors and scalar-spinor interaction terms. Moreover, this theory still presents a remarkable equivalence between the Noether and topological currents, describes the soliton sector of the original model and turns out to be the master Lagrangian describing the GMT and GSG models.

1 Introduction

Integrable theories in two-dimensions have been an extraordinary laboratory for the understanding of basic nonperturbative aspects of physical theories and various aspects, relevant in more realistic 4-dimensional models, have been tested [1]. In particular the conformal affine Toda models coupled to (Dirac) matter fields (CATM) [2] for the $sl(2)^{(1)}$ and $sl(3)^{(1)}$ cases are discussed in [3, 4, 5] and [6, 7, 8], respectively. The interest in such models comes from their integrability and duality properties [2, 4], which can be used as toy models to understand some phenomena; such as, a confinement mechanism in QCD [3, 6, 9]. The corresponding off-critical sub-models, such as the $\hat{sl}(2)$ affine Toda model coupled to matter field (ATM) may describe some low dimensional condensed matter phenomena, such as self-trapping of electrons into solitons, see e.g. [10], tunneling in the integer quantum Hall effect [11], and, in particular, polyacteline molecule systems in connection with fermion number fractionization [12].

The off-critical $\hat{sl}(2)$ ATM model can be obtained at the classical or quantum mechanical level through some convenient reduction processes starting from CATM [4, 5]. In the $sl(2)$ case, using bosonization techniques, it has been shown that the classical equivalence between the $U(1)$ vector and topological currents holds true at the quantum level, and then leads to a bag model like mechanism for the confinement of the spinor fields inside the solitons; in addition, it has been shown that the $sl(2)$ ATM theory decouples into a sine-Gordon model (SG) and a free scalar [3, 13]. These facts indicate the existence of a sort of duality in these models involving solitons and particles. In ref. [5] through the Faddeev-Jackiw (FJ)[14] and "symplectic quantization" [15, 16] methods and imposing the equivalence between the $U(1)$ vector and topological currents as a constraint it has been recovered the usual massive Thirring (MT) and sine-Gordon (SG) models equivalence at the classical level; in particular, the mappings between spinor bilinears of the MT theory and exponentials of the SG fields were established on shell and the soliton/particle correspondence were uncovered [4]. Moreover, it has recently been shown that the $\hat{sl}(2)$ ATM model describes a confinement mechanism and the low-energy spectrum of QCD_2 (one flavor and N colors) [9].

One of the difficulties with generalizations of complex affine Toda field theories, beyond $su(2)$ and its associated SG model, has to do with unitarity. In this chapter we describe the many field generalizations of SG/MT models based on soliton/particle duality and unitarity. Beyond the well known $sl(2)$ case the related $sl(n)^{(1)}$ CATM model does not possess a local real Lagrangian, therefore it is defined an off-critical sub-model Lagrangian with well behaved classical solutions making use of the results of [6]. In [6] the authors studied the $sl(3)^{(1)}$ CATM soliton solutions and some of their properties up to general 2-soliton. Using the FJ and symplectic methods it has been show the parent Lagrangian [17] nature of the $sl(3)$ ATM model from which the generalized sine-Gordon (GSG) or the massive Thirring (GMT) models are derivable [7, 8]. It is then shown that there are (at least classically) two equivalent descriptions of the model, by means of either the Dirac or the Toda type fields. It is also clear the duality exchange of the coupling regimes $g \to 1/g$ in each $\hat{sl}(2)$ ATM submodel, in this way generalizing the soliton/particle correspondences which is uncovered by providing explicit relationships between the GSG and GMT fields. The ordinary MT/SG duality turns out to be related to each $SU(2)$ sub-group. The $\hat{sl}(n)$ affine Lie algebra generalizations are outlined following [8]. In this way we describe a precise field content of both sectors; namely, the correct GMT/GSG duality, first undertaken in [18].

Recently, the three fermion species GMT model has been bosonized and the $\hat{sl}(3)$ ATM model emerged in the semi-classical limit of an intermediate Lagrangian [19]. The generalized Mandelstam soliton operators were constructed and the fermion-boson mapping was established through a set of generalized bosonization rules in a quotient positive definite Hilbert space of states.

The chapter is organized as follows. In section 2 we define the $\hat{sl}(3)$ ATM model . Section 3 deals with the model in the FJ framework [14], the outcome is the GMT model. In section 4, we attack the same problem from the point of view of symplectic quantization [15, 16] giving the Poisson brackets of the GMT and GSG models. Section 5 deals with the soliton/particle and strong/weak coupling correspondences. Section 6 provides a full Lie algebraic construction of the relevant results. Section 7 outlines the relevant steps towards the generalization to $\hat{sl}(n)$ ATM. In the appendix A we present the construction of $sl(3)^{(1)}$ CATM model.

2 The model

In affine Toda type theories the question of whether all mathematical solutions are physically acceptable deserves a careful analysis, specially if any consistent quantization of the models is discussed. The requirement of real energy density leads to a certain reality conditions on the solutions of the model. In general, a few soliton solutions survive the reality constraint, if in addition one also demands positivity. These kind of issues are discussed in refs. [20]. Here we follow the prescription to restrict the model to a subspace of classical solutions which satisfy the physical principles of reality of energy density and soliton/particle correspondence.

In CATM models associated to the principal gradation of an affine Lie algebra, it is a matter of fact that a 1-soliton real solution for the Toda field corresponds to each pair of Dirac fields ψ^i and $\widetilde{\psi}^i$. This fact allows us to make the identifications $\widetilde{\psi}^i \sim (\psi^i)^*$($\star$ means complex conjugation), and take real Toda fields. In the case of $sl(2)^{(1)}$ CATM theory, this procedure does not spoil the particle-soliton correspondence [3, 4].

We consider the $sl(3)^{(1)}$ CATM theory (see Appendix A) with the conformal symmetry gauge fixed [1] by setting $\eta = 0$ and the reality conditions

$$\widetilde{\psi}^j = -(\psi^j)^*, \; (j=1,2,3); \;\; \varphi_a^* = \varphi_a, \; (a=1,2), \tag{1}$$

or

$$\begin{aligned} \widetilde{\psi}^j &= (\psi^j)^*, \; j=1,2; \quad \widetilde{\psi}^3 = -(\psi^3)^*, \\ \varphi_{1,2} &\rightarrow \varphi_{1,2} - \pi \quad \text{(the new } \varphi_a\text{'s being real fields)}. \end{aligned} \tag{2}$$

The condition (2) must be supplied with $x^\mu \rightarrow -x^\mu$. Moreover, for consistency of the equations of motion (111)-(119) under the reality conditions (1)-(2), from eqs. (112), (113), (114), (116), (117) and (119), we get the relationships

$$\widetilde{\psi}_L^j \psi_R^3 - \widetilde{\psi}_R^j \psi_L^3 e^{-3i\varphi_j} = 0, \; j=1,2; \;\; \psi_L^1 \psi_R^2 e^{-3i\varphi_1} - \psi_L^2 \psi_R^1 e^{-3i\varphi_2} = 0. \tag{3}$$

Then, the above reality conditions and constraints allow us to define a suitable physical Lagrangian. The equations (109), (111)-(119), supplied with (1) [or (2)] and (3), follow

[1] The auxiliary fields $\widetilde{\nu}$ and η of the CATM theories are associated to the topological character of the soliton masses and to the conformal symmetry, respectively. The classical and quantum reductions CATM $\rightarrow$ ATM can be treated as in [5] and [4], respectively.

from the Lagrangian

$$\frac{1}{k}\mathcal{L} = \sum_{j=1}^{3}\left[\frac{1}{24}\partial_\mu\phi_j\partial^\mu\phi_j + i\overline{\psi}^j\gamma^\mu\partial_\mu\psi^j - m_\psi^j\overline{\psi}^j e^{i\phi_j\gamma_5}\psi^j\right] \tag{4}$$

where $\bar{\psi}^j \equiv (\psi^j)^\dagger\,\gamma_0$, $\phi_1 \equiv 2\varphi_1 - \varphi_2$, $\phi_2 \equiv 2\varphi_2 - \varphi_1$, $\phi_3 \equiv \phi_1 + \phi_2$, $m_\psi^3 = m_\psi^1 + m_\psi^2$, k is an overall coupling constant and the φ_j are real fields.

The eq. (4) defines the $sl(3)$ *affine Toda theory coupled to matter fields* (ATM). Notice that the space of solutions of $sl(3)^{(1)}$ CATM model satisfying the conditions (1)-(3) must be solutions of the $sl(3)$ ATM theory (4). Indeed, it is easy to verify that the three species of one-soliton solutions [$S \equiv$*1-soliton*($\bar{S} \equiv$*1-antisoliton*)] [6]: $\{\left(\varphi_1,\,\psi^1\right)_{S/\bar{S}},\,\varphi_2 = 0,\,\psi^2 = 0,\,\psi^3 = 0\}$, $\{\left(\varphi_2,\,\psi^2\right)_{S/\bar{S}},\,\varphi_1 = 0,\,\psi^1 = 0,\,\psi^3 = 0\}$ and $\{\left(\varphi_1+\varphi_2,\,\psi^3\right)_{S/\bar{S}},\,\varphi_1 = \varphi_2,\,\psi^1 = 0,\,\psi^2 = 0\}$ satisfy the equations of motion; i.e., each positive root of $sl(3)$ reproduces the $sl(2)$ ATM case [3, 4]. Moreover, these solutions satisfy the above reality conditions and constraints (1)-(3) (with (1) and (2) for S and $\bar{S}$, respectively), and the equivalence between the $U(1)$ vector and topological currents (125). Then, the soliton/particle correspondences survive the above reduction processes performed to define the $sl(3)$ ATM theory.

The class of 2-soliton solutions of $sl(3)^{(1)}$ CATM [6] behave as follows: i) they are given by 6 species associated to the pair (α_i, α_j), $i \leq j$; $i, j = 1, 2, 3$; where the α's are the positive roots of $sl(3)$ Lie algebra. Each species (α_i, α_i) solves the $sl(2)$ CATM submodel[2]; ii) satisfy the $U(1)$ vector and topological currents equivalence (125).

3 The Generalized Massive Thirring Model (GMT)

Let us consider the following Lagrangian

$$\frac{1}{k}\mathcal{L} = \sum_{j=1}^{3}\left[\frac{1}{24}\partial_\mu\phi_j\partial^\mu\phi_j + i\overline{\psi}^j\gamma^\mu\partial_\mu\psi^j - m_\psi^j\overline{\psi}^j e^{i\phi_j\gamma_5}\psi^j + \lambda_\mu^j(m^j\overline{\psi}^j\gamma^\mu\psi^j - \epsilon^{\mu\nu}\partial_\nu(q_j\phi_j))\right], \tag{5}$$

where the ATM Lagrangian (4) is supplied with the constraints, $(m^l\overline{\psi}^l\gamma^\mu\psi^l + \frac{m^3}{2}\overline{\psi}^3\gamma^\mu\psi^3 - \epsilon^{\mu\nu}\partial_\nu\phi_l)$, $(l = 1, 2)$, with the help of the Lagrange multipliers λ_μ^j $(\lambda_\mu^3 \equiv \frac{\lambda_\mu^1+\lambda_\mu^2}{2}$, $q_1 \equiv q_2 \equiv 1,\ q_3 \equiv 0)$. Their total sum bears an intriguing resemblance to the $U(1)$ vector and topological currents equivalence (125); however, the m^j's here are some arbitrary parameters. The same procedure has been used, for example, to incorporate the left-moving condition in the study of chiral bosons in two dimensions [21]. The Lagrangian (5) still possesses the left-right local symmetries (121)-(124) of $sl(3)$ ATM (4) provided a convenient transformation for the field λ_j^μ is assumed[3]. In order to apply the Faddeev-Jackiw (FJ)

[2] $sl(2)$ ATM 2−solitons satisfy an analogous eq. to (125); for φ real and $\widetilde{\psi} = \pm(\psi)^*$ (constraints (1)-(2); (3) is trivialy satisfied since $\widetilde{\psi}_j = \psi_j = 0$ for $j \neq i$) one has, soliton-soliton SS, SS bounds and no $S\bar{S}$ bounds [3].

[3] In [7] this feature has not been pointed out.

method we should write (5) in the first order form in time derivative, so let us define the conjugated momenta

$$\begin{aligned}\pi_1 &\equiv \pi_{\phi_1} = \frac{1}{12}(2\dot{\phi}_1 + \dot{\phi}_2) + \lambda_1^1, \qquad \pi_2 \equiv \pi_{\phi_2} = \frac{1}{12}(2\dot{\phi}_2 + \dot{\phi}_1) + \lambda_1^2,\\ \pi_{\lambda_\mu^1} &= 0, \qquad \pi_{\lambda_\mu^2} = 0, \qquad \pi_R^j \equiv \pi_{\psi_R}^j = -i\tilde{\psi}_R^j, \qquad \pi_L^j \equiv \pi_{\psi_L}^j = -i\tilde{\psi}_L^j. \end{aligned} \tag{6}$$

We are assuming that Dirac fields are anticommuting Grasmannian variables and their momenta variables defined through **left** derivatives. Then, as usual, the Hamiltonian is defined by (sum over repeated indices is assumed)

$$\mathcal{H}_c = \pi_1\dot{\phi}_1 + \pi_2\dot{\phi}_2 + \dot{\psi}_R^j\pi_R^j + \dot{\psi}_L^j\pi_L^j - \mathcal{L}. \tag{7}$$

Explicitly the Hamiltonian density becomes

$$\begin{aligned}\mathcal{H}_c &= 2(\pi_j)^2 + 4(\lambda_1^1)^2 + 4\left(\lambda_1^2\right)^2 - \lambda_1^1\mathcal{J}_1 - \lambda_1^2\mathcal{J}_2 - 4\left(\lambda_1^1\lambda_1^2\right)\\ &+\frac{1}{24}(\phi_{j,x})^2 - \pi_R^j\psi_{R,x}^j + \pi_L^j\psi_{L,x}^j + im_\psi^j(e^{-\phi_j}\tilde{\psi}_R^j\psi_L^j - e^{\phi_j}\tilde{\psi}_L^j\psi_R^j)\\ &+\lambda_0^1[J_1^0 - \phi_{1,x}] + \lambda_0^2[J_2^0 - \phi_{2,x}],\end{aligned} \tag{8}$$

where $\pi_3 \equiv \pi_1 - \pi_2$, $\mathcal{J}_1 \equiv J_1^1 + 4(2\pi_1 - \pi_2)$, $\mathcal{J}_2 \equiv J_2^1 + 4(2\pi_2 - \pi_1)$ and

$$J_1^\mu = m^1 j_l^\mu + \frac{m^3}{2}j_3^\mu; \quad J_2^\mu = m^2 j_2^\mu + \frac{m^3}{2}j_3^\mu; \quad j_l^\mu \equiv \bar{\psi}^l\gamma^\mu\psi^l, \quad l = 1,2,3. \tag{9}$$

Let us observe that each $U(1)$ Noether current of the $sl(3)$ ATM theory defined in (4) is conserved separately; i.e., $\partial_\mu j_l^\mu = 0, \quad l = 1,2,3.$

Next, the same Legendre transform (7) is used to write the first order Lagrangian

$$\mathcal{L} = \pi_1\dot{\phi}_1 + \pi_2\dot{\phi}_2 + \dot{\psi}_R^j\pi_R^j + \dot{\psi}_L^j\pi_L^j - \mathcal{H}_c. \tag{10}$$

Our starting point for the FJ analysis will be this first order Lagrangian. Then the Euler-Lagrange equations for the Lagrange multipliers allow one to solve two of them

$$\lambda_1^1 = \frac{2\mathcal{J}_1 + \mathcal{J}_2}{12}, \qquad \lambda_1^2 = \frac{2\mathcal{J}_2 + \mathcal{J}_1}{12} \tag{11}$$

and the remaining equations lead to two constraints

$$\Omega_1 \equiv J_1^0 - \phi_{1,x} = 0, \qquad \Omega_2 \equiv J_2^0 - \phi_{2,x} = 0. \tag{12}$$

The Lagrange multipliers λ_1^1 and λ_1^2 must be replaced back in (10) and the constraints (12) solved. Firstly, let us replace the λ_1^1 and λ_1^2 multipliers into $\mathcal{H}_c$, then one gets

$$\begin{aligned}\mathcal{H}_c' &= 2(\pi_j)^2 - \frac{1}{12}\{(\mathcal{J}_1)^2 + (\mathcal{J}_2)^2 + (\mathcal{J}_1\mathcal{J}_2)\} + \frac{1}{24}(\phi_{j,x})^2\\ &+ i\tilde{\psi}_R^j\psi_{R,x}^j - i\tilde{\psi}_L^j\psi_{L,x}^j + im_\psi^j(e^{-i\phi_j}\tilde{\psi}_R^j\psi_L^j - e^{i\phi_j}\tilde{\psi}_L^j\psi_R^j).\end{aligned} \tag{13}$$

The new Lagrangian becomes

$$\mathcal{L}' = \pi_1\dot{\phi}_1 + \pi_2\dot{\phi}_2 + \dot{\psi}_R^j\pi_R^j + \dot{\psi}_L^j\pi_L^j - \mathcal{H}_c'. \tag{14}$$

We implement the constraints (12) by replacing in (14) the fields ϕ_1, ϕ_2 in terms of the space integral of the current components J_1^0, J_2^0. Then we get the Lagrangian

$$\begin{aligned} \mathcal{L}'' &= \pi_1 \partial_t \int^x J_1^0 + \pi_2 \partial_t \int^x J_2^0 + \dot{\psi}_R^j \pi_R^j + \dot{\psi}_L^j \pi_L^j - i\tilde{\psi}_R^j \psi_{R,x}^j + i\tilde{\psi}_L^j \psi_{L,x}^j \\ &- i m_\psi^j \Big(e^{-i\int^x J_j^0} \tilde{\psi}_R^j \psi_L^j - e^{i\int^x J_j^0} \tilde{\psi}_L^j \psi_R^j \Big) - \frac{1}{12}\Big((J_1)^2 + (J_2)^2 + J_1 . J_2 \Big) \\ &+ \pi_1 J_1^1 + \pi_2 J_2^1, \end{aligned} \tag{15}$$

where $J_3^0 \equiv J_1^0 + J_2^0$. Observe that the terms containing the π_a's in eq. (15) cancel to each other if one uses the current conservation laws. Notice the appearances of various types of current-current interactions. The following Darboux transformation

$$\psi_R^j \rightarrow e^{-\frac{i}{2}\int^x J_j^0} \psi_R^j, \qquad \psi_L^j \rightarrow e^{\frac{i}{2}\int^x J_j^0} \psi_L^j, \quad j = 1, 2, 3, \tag{16}$$

is used to diagonalize the canonical one-form. Then, the kinetic terms will give additional current-current interactions, $-\frac{1}{2}[J_1.(j_1 + j_3) + J_2.(j_2 + j_3)]$. We are, thus, after defining $k \equiv 1/g$, and rescaling the fields $\psi^j \rightarrow 1/\sqrt{k}\, \psi^j$, left with the Lagrangian

$$\mathcal{L}[\psi, \overline{\psi}] = \sum_{j=1}^{3} \{ i\overline{\psi}^j \gamma^\mu \partial_\mu \psi^j + m_\psi^j \, \overline{\psi}^j \psi^j \} - \sum_{\substack{k,l=1 \\ k \le l}}^{3} \Big[\bar{a}_{kl} j_k . j_l \Big], \tag{17}$$

where $\bar{a}_{kl} = g\, a_{kl}$, with $a_{33} = \frac{1}{2}(\frac{(m^3)^2}{8} + m^3)$; $a_{12} = \frac{1}{12} m^1 m^2$; $a_{ii} = \frac{1}{2}(\frac{(m^i)^2}{6} + m^i)$; $a_{i3} = \frac{1}{2}(\frac{m^i m^3}{4} + m^i + \frac{m^3}{2})$, $i = 1, 2$. This defines the *generalized massive Thirring model* (GMT) [7, 8]. The canonical pairs are $(-i\tilde{\psi}_R^j, \psi_R^j)$ and $(-i\tilde{\psi}_L^j, \psi_L^j)$. Let us comment on the integrability and the Grassmannian nature of the ψ fields. The usual MT model (with Grassmannian fields) classical integrability has been established by means of its zero-curvature formulation (see e.g. [22]). In our case it would be interesting to establish the GMT integrability using similar approach.

From the Lie algebraic constructions of [8, 19] one has all $\bar{a}_{jk} > 0$, $j, k = 1, 2, 3$; but $\bar{a}_{12} < 0$. Taking into account the *signs* of the $\bar{a}_{jk}$'s in the model (17) one can infer that the fermions of the same species will experience an attractive force. The pair of fermions of species 1 and 3, as well as 2 and 3 also experience attractive forces, whereas the pair of fermions 1 and 2 suffer a repulsive force [23]. These features can also be deduced from the behavior of the time delays due to soliton-soliton interactions in the associated $\hat{su}(3)$ ATM model studied in ref. [6].

4 The Symplectic Formalism and the ATM Model

4.1 The (Constrained) Symplectic Formalism

We give a brief overview of the basic notations of symplectic approach [4]. The geometric structure is defined by the closed (pre)symplectic two-form

$$f^{(0)} = \frac{1}{2} f_{ij}^{(0)}(\xi^{(0)}) d\xi^{(0)^i} \wedge d\xi^{(0)^j} \tag{18}$$

[4]These are given for point mechanics, the extension to field theory is self evident.

where

$$f_{ij}^{(0)}(\xi^{(0)}) = \frac{\partial}{\partial \xi^{(0)^i}} \mathbf{a}_j^{(0)}(\xi^{(0)}) - \frac{\partial}{\partial \xi^{(0)^j}} \mathbf{a}_i^{(0)}(\xi^{(0)}) \tag{19}$$

with $\mathbf{a}^{(0)}(\xi^{(0)}) = \mathbf{a}_j^{(0)}(\xi^{(0)}) d\xi^{(0)^j}$ being the canonical one-form defined from the original first order Lagrangian

$$L^{(0)} dt = \mathbf{a}^{(0)}(\xi^{(0)}) - V^{(0)}(\xi^{(0)}) dt. \tag{20}$$

The superscript (0) refers to the original Lagrangian, and is indicative of the iterative nature of the computations. The constraints are imposed through Lagrange multipliers which are velocities, and in such case one has to extend the configuration space [15, 16]. The corresponding Lagrangian gets modified and consequently the superscript also changes. The algorithm terminates once the symplectic matrix turns out to be non-singular.

4.2 The Generalized Massive Thirring Model (GMT)

Next, we will consider our model in the framework of the symplectic formalism. Let $\mathcal{L}'$, eq. (14), be the zeroth-iterated Lagrangian $\mathcal{L}^{(0)}$. Then the first iterated lagrangian will be

$$\mathcal{L}^{(1)} - \pi_1 \dot{\phi}_1 + \pi_2 \dot{\phi}_2 + \dot{\psi}_R^j \pi_R^j + \dot{\psi}_L^j \pi_L^j + \dot{\eta}^1 \Omega_1 + \dot{\eta}^2 \Omega_2 - \mathcal{V}^{(1)}, \tag{21}$$

where the once-iterated symplectic potential is defined by

$$\mathcal{V}^{(1)} = \mathcal{H}_c' |_{\Omega_1 = \Omega_2 = 0}, \tag{22}$$

and the stability conditions of the symplectic constraints, Ω_1 and Ω_2, under time evolution have been implemented by making $\lambda_0^1 \to \dot{\eta}^1$ and $\lambda_0^2 \to \dot{\eta}^2$. Consider the once-iterated set of symplectic variables in the following order

$$\xi^{(1)} = (\eta^1, \eta^2, \phi_1, \phi_2, \psi_R^1, \psi_L^1, \psi_R^2, \psi_L^2, \psi_R^3, \psi_L^3, \pi_1, \pi_2, \pi_R^1, \pi_L^1, \pi_R^2, \pi_L^2, \pi_R^3, \pi_L^3), \tag{23}$$

and the components of the canonical one-form

$$\mathbf{a}^{(1)} = (\Omega_1, \Omega_2, \pi_1, \pi_2, -\pi_R^1, -\pi_L^1, -\pi_R^2, -\pi_L^2, -\pi_R^3, -\pi_L^3, 0, 0, 0, 0, 0, 0, 0, 0). \tag{24}$$

These result in the singular symplectic two-form 18x18 matrix

$$f_{AB}^{(1)}(x, y) = \begin{pmatrix} a_{11} & a_{12} \\ a_{21} & a_{22} \end{pmatrix} \delta(x - y), \tag{25}$$

where the 9x9 matrices are

$$a_{11} = \begin{pmatrix} 0 & 0 & \partial_x & 0 & im^1 \tilde{\psi}_R^1 & im^1 \tilde{\psi}_L^1 & 0 & 0 & \frac{im^3}{2} \tilde{\psi}_R^3 \\ 0 & 0 & 0 & \partial_x & 0 & 0 & im^2 \tilde{\psi}_R^2 & im^2 \tilde{\psi}_L^2 & \frac{im^3}{2} \tilde{\psi}_R^3 \\ \partial_x & 0 & 0 & 0 & 0 & 0 & 0 & 0 & 0 \\ 0 & \partial_x & 0 & 0 & 0 & 0 & 0 & 0 & 0 \\ im^1 \tilde{\psi}_R^1 & 0 & 0 & 0 & 0 & 0 & 0 & 0 & 0 \\ im^1 \tilde{\psi}_L^1 & 0 & 0 & 0 & 0 & 0 & 0 & 0 & 0 \\ 0 & im^2 \tilde{\psi}_R^2 & 0 & 0 & 0 & 0 & 0 & 0 & 0 \\ 0 & im^2 \tilde{\psi}_L^2 & 0 & 0 & 0 & 0 & 0 & 0 & 0 \\ \frac{im^3}{2} \tilde{\psi}_R^3 & \frac{im^3}{2} \tilde{\psi}_R^3 & 0 & 0 & 0 & 0 & 0 & 0 & 0 \end{pmatrix},$$

$$a_{12} = \begin{pmatrix} \frac{im^3}{2}\tilde{\psi}^3_L & 0 & 0 & m^1\psi^1_R & m^1\psi^1_L & 0 & 0 & \frac{m^3}{2}\psi^3_R & \frac{m^3}{2}\psi^3_L \\ \frac{im^3}{2}\tilde{\psi}^3_L & 0 & 0 & 0 & 0 & m^2\psi^2_R & m^2\psi^2_L & \frac{m^3}{2}\psi^3_R & \frac{m^3}{2}\psi^3_L \\ 0 & -1 & 0 & 0 & 0 & 0 & 0 & 0 & 0 \\ \vdots & 0 & \ddots & \vdots & \vdots & \vdots & \vdots & \vdots & \vdots \\ 0 & 0 & 0 & 0 & 0 & 0 & 0 & -1 & 0 \end{pmatrix},$$

$$a_{21} = \begin{pmatrix} \frac{im^3}{2}\tilde{\psi}^3_L & \frac{im^3}{2}\tilde{\psi}^3_L & 0 & 0 & 0 & 0 & 0 & 0 & 0 \\ 0 & 0 & 1 & 0 & 0 & 0 & 0 & 0 & 0 \\ 0 & 0 & 0 & 1 & 0 & 0 & 0 & 0 & 0 \\ m^1\psi^1_R & 0 & 0 & 0 & -1 & 0 & 0 & 0 & 0 \\ m^1\psi^1_L & 0 & 0 & 0 & 0 & -1 & 0 & 0 & 0 \\ 0 & m^2\psi^2_R & 0 & 0 & 0 & 0 & -1 & 0 & 0 \\ 0 & m^2\psi^2_L & 0 & 0 & 0 & 0 & 0 & -1 & 0 \\ \frac{m^3}{2}\psi^3_R & \frac{m^3}{2}\psi^3_R & 0 & 0 & 0 & 0 & 0 & 0 & -1 \\ \frac{m^3}{2}\psi^3_L & \frac{m^3}{2}\psi^3_L & 0 & 0 & 0 & 0 & 0 & 0 & 0 \end{pmatrix},$$

$$a_{22} = \begin{pmatrix} 0 & 0 & \cdots & 0 & -1 \\ 0 & 0 & 0 & 0 & 0 \\ \cdots & \cdots & \cdots & \cdots & \cdots \\ 0 & 0 & 0 & 0 & 0 \\ -1 & 0 & \cdots & 0 & 0 \end{pmatrix}.$$

This matrix has the zero modes

$$\begin{aligned} \mathbf{v}^{(1)^T}(x) = & \Big(\frac{-u}{m^1}, \frac{-v}{m^2}, 0, 0, u\psi^1_R, u\psi^1_L, v\psi^2_R, v\psi^2_L, \frac{m^3}{2}\Big(\frac{u}{m^1}+\frac{v}{m^2}\Big)\psi^3_R, \\ & \frac{m^3}{2}\Big(\frac{u}{m^1}+\frac{v}{m^2}\Big)\psi^3_L, -\frac{u'}{m^1}, -\frac{v'}{m^2}, iu\tilde{\psi}^1_R, iu\tilde{\psi}^1_L, iv\tilde{\psi}^2_R, iv\tilde{\psi}^2_L, \\ & i\frac{m^3}{2}(\frac{u}{m^1}+\frac{v}{m^2})\tilde{\psi}^3_R, \frac{im^3}{2}(\frac{u}{m^1}+\frac{v}{m^2})\tilde{\psi}^3_L\Big), \end{aligned} \tag{26}$$

where u and v are arbitrary functions. The zero-mode condition gives

$$\int dx \mathbf{v}^{(1)^T}(x)\frac{\delta}{\delta\xi^{(1)}(x)}\int dy \mathcal{V}^{(1)} \equiv 0. \tag{27}$$

Thus, the gradient of the symplectic potential happens to be orthogonal to the zero-mode $\mathbf{v}^{(1)}$. Since the equations of motion are automatically validated no symplectic constraints appear. This happens due to the presence of the symmetries of the action

$$\delta\xi^{(1)}_A = \mathbf{v}^{(1)}_{\mathbf{A}}(x); \quad A = 1, 2, ...18. \tag{28}$$

So, in order to deform the symplectic matrix into an invertible one, we have to add some gauge fixing terms to the symplectic potential. One can choose any consistent set of gauge fixing conditions [16]. In our case we have two symmetry generators associated to the parameters u and v, so there must be two gauge conditions. Let us choose

$$\Omega_3 \equiv \phi_1 = 0 \qquad \Omega_4 \equiv \phi_2 = 0. \tag{29}$$

These conditions gauge away the fields ϕ_1 and ϕ_2, so only the remaining field variables will describe the dynamics of the system. Other gauge conditions, which eventually gauge away the spinor fields ψ^i will be considered in the next subsection.

Implementing the consistency conditions by means of Lagrange multipliers η^3 and η^4 we get the twice-iterated Lagrangian

$$\mathcal{L}^{(2)} = \pi_1\dot{\phi}_1 + \pi_2\dot{\phi}_2 + \dot{\psi}_R\pi_R^j + \dot{\psi}_L\pi_L^j + \dot{\eta}^1\Omega_1 + \dot{\eta}^2\Omega_2 + \dot{\eta}^3\Omega_3 + \dot{\eta}^4\Omega_4 - \mathcal{V}^{(2)}, \tag{30}$$

where

$$\mathcal{V}^{(2)} = \mathcal{V}^{(1)}|_{\Omega_3=\Omega_4=0}.$$

Assuming now that the new set of symplectic variables is given in the following order

$$\xi^{(2)} = (\eta^1, \eta^2, \eta^3, \eta^4, \phi_1, \phi_2, \psi_R^1, \psi_L^1, \psi_R^2, \psi_L^2, \psi_R^3, \psi_L^3, \pi_1, \pi_2, \pi_R^1, \pi_L^1, \pi_R^2, \pi_L^2, \pi_R^3, \pi_L^3), \tag{31}$$

and the non vanishing components of the canonical one-form

$$\mathbf{a}^{(2)} = (\Omega_1, \Omega_2, \Omega_3, \Omega_4, \pi_1, \pi_2, -\pi_R^1, -\pi_L^1, -\pi_R^2, -\pi_L^2, -\pi_R^3, -\pi_L^3, 0, 0, 0, 0, 0, 0, 0, 0), \tag{32}$$

one obtains the singular twice-iterated symplectic 20x20 matrix

$$f_{AB}^{(2)}(x,y) = \begin{pmatrix} a_{11} & a_{12} \\ a_{21} & a_{22} \end{pmatrix} \delta(x-y), \tag{33}$$

where the 10x10 matrices are

$$a_{11} = \begin{pmatrix} 0 & 0 & 0 & 0 & \partial_x & 0 & im^1\tilde{\psi}_R^1 & im^1\tilde{\psi}_L^1 & 0 & 0 \\ 0 & 0 & 0 & 0 & 0 & \partial_x & 0 & 0 & im^2\tilde{\psi}_R^2 & im^2\tilde{\psi}_L^2 \\ 0 & 0 & 0 & 0 & -1 & 0 & 0 & 0 & 0 & 0 \\ 0 & 0 & 0 & 0 & 0 & -1 & 0 & 0 & 0 & 0 \\ \partial_x & 0 & 1 & 0 & 0 & 0 & 0 & 0 & 0 & 0 \\ 0 & \partial_x & 0 & 1 & 0 & 0 & 0 & 0 & 0 & 0 \\ im^1\tilde{\psi}_R^1 & 0 & 0 & 0 & 0 & 0 & 0 & 0 & 0 & 0 \\ im^1\tilde{\psi}_L^1 & 0 & 0 & 0 & 0 & 0 & 0 & 0 & 0 & 0 \\ 0 & im^2\tilde{\psi}_R^2 & 0 & 0 & 0 & 0 & 0 & 0 & 0 & 0 \\ 0 & im^2\tilde{\psi}_L^2 & 0 & 0 & 0 & 0 & 0 & 0 & 0 & 0 \end{pmatrix},$$

$$a_{12} = \begin{pmatrix} \frac{im^3}{2}\tilde{\psi}_R^3 & \frac{im^3}{2}\tilde{\psi}_L^3 & 0 & 0 & m^1\psi_R^1 & m^1\psi_L^1 & 0 & 0 & \frac{m^3}{2}\psi_R^3 & \frac{m^3}{2}\psi_L^3 \\ \frac{im^3}{2}\tilde{\psi}_R^3 & \frac{im^3}{2}\tilde{\psi}_L^3 & 0 & 0 & 0 & 0 & m^2\psi_R^2 & m^2\psi_L^2 & \frac{m^3}{2}\psi_R^3 & \frac{m^3}{2}\psi_L^3 \\ 0 & 0 & 0 & 0 & 0 & 0 & 0 & 0 & 0 & 0 \\ 0 & 0 & 0 & 0 & 0 & 0 & 0 & 0 & 0 & 0 \\ 0 & 0 & -1 & 0 & 0 & 0 & 0 & 0 & 0 & 0 \\ 0 & 0 & 0 & -1 & 0 & 0 & 0 & 0 & 0 & 0 \\ 0 & 0 & 0 & 0 & -1 & 0 & 0 & 0 & 0 & 0 \\ 0 & 0 & 0 & 0 & 0 & -1 & 0 & 0 & 0 & 0 \\ 0 & 0 & 0 & 0 & 0 & 0 & -1 & 0 & 0 & 0 \\ 0 & 0 & 0 & 0 & 0 & 0 & 0 & -1 & 0 & 0 \end{pmatrix},$$

$$a_{21} = \begin{pmatrix} \frac{im^3}{2}\tilde{\psi}_R^3 & \frac{im^3}{2}\tilde{\psi}_R^3 & 0 & 0 & 0 & 0 & 0 & 0 & 0 & 0 \\ \frac{im^3}{2}\tilde{\psi}_L^3 & \frac{im^3}{2}\tilde{\psi}_L^3 & 0 & 0 & 0 & 0 & 0 & 0 & 0 & 0 \\ 0 & 0 & 0 & 0 & 1 & 0 & 0 & 0 & 0 & 0 \\ 0 & 0 & 0 & 0 & 0 & 1 & 0 & 0 & 0 & 0 \\ m^1\psi_R^1 & 0 & 0 & 0 & 0 & 0 & -1 & 0 & 0 & 0 \\ m^1\psi_L^1 & 0 & 0 & 0 & 0 & 0 & 0 & -1 & 0 & 0 \\ 0 & m^2\psi_R^2 & 0 & 0 & 0 & 0 & 0 & 0 & -1 & 0 \\ 0 & m^2\psi_L^2 & 0 & 0 & 0 & 0 & 0 & 0 & 0 & -1 \\ \frac{m^3}{2}\psi_R^3 & \frac{m^3}{2}\psi_R^3 & 0 & 0 & 0 & 0 & 0 & 0 & 0 & 0 \\ \frac{m^3}{2}\psi_L^3 & \frac{m^3}{2}\psi_L^3 & 0 & 0 & 0 & 0 & 0 & 0 & 0 & 0 \end{pmatrix},$$

$$a_{22} = \begin{pmatrix} 0 & 0 & 0 & \cdots & 0 & -1 & 0 \\ 0 & 0 & 0 & \cdots & 0 & 0 & -1 \\ 0 & 0 & 0 & \cdots & 0 & 0 & 0 \\ \cdots & & & & & & \cdots \\ 0 & 0 & 0 & \cdots & 0 & 0 & 0 \\ -1 & 0 & 0 & \cdots & 0 & 0 & 0 \\ 0 & -1 & 0 & \cdots & 0 & 0 & 0 \end{pmatrix}.$$

The zero-modes are

$$\begin{aligned} \mathbf{v}^{(2)^T}(x) &= \Big(u,\ v,\ \omega,\ \chi,\ 0,\ 0,\ m^1u\psi_R^1,\ m^1u\psi_L^1,\ m^2v\psi_R^2,\ m^2v\psi_L^2,\ \frac{m^3}{2}(u+v)\psi_R^3, \\ &\quad \frac{m^3}{2}(u+v)\psi_L^3,\ u'+\omega,\ v'+\chi,\ im^1\tilde{\psi}_R^1u,\ im^1\tilde{\psi}_L^1u,\ im^2\tilde{\psi}_R^2v, \\ &\quad im^2\tilde{\psi}_L^2v,\ \frac{im^3}{2}\tilde{\psi}_R^3(u+v),\ i\frac{m^3}{2}\tilde{\psi}_L^3(u+v)\Big). \end{aligned} \tag{34}$$

The zero-mode condition gives no constraints, implying the symmetries of the action

$$\delta\xi_{\mathbf{A}}^{(2)} = \mathbf{v}_{\mathbf{A}}^{(2)}(x); \quad A = 1, 2, ...20. \tag{35}$$

Now, let us choose the gauge conditions

$$\Omega_5 \equiv \pi_1 J_1^1 + \frac{1}{2}J_1.(j_1+j_3) = 0, \qquad \Omega_6 \equiv \pi_2 J_2^1 + \frac{1}{2}J_2.(j_2+j_3) = 0, \tag{36}$$

and impose the consistency conditions with the Lagrange multipliers η^5, η^6, then

$$\mathcal{L}^{(3)} = \pi_1\dot{\phi}_1 + \pi_2\dot{\phi}_2 + \dot{\psi}_R\pi_R^j + \dot{\psi}_L\pi_L^j + \dot{\eta}^1\Omega_1 + \dot{\eta}^2\Omega_2 + \dot{\eta}^3\Omega_3 + \dot{\eta}^4\Omega_4 + \dot{\eta}^5\Omega_5 + \dot{\eta}^6\Omega_6 - \mathcal{V}^{(3)}, \tag{37}$$

where

$$\mathcal{V}^{(3)} = \mathcal{V}^{(2)}|_{\Omega_5=\Omega_6=0}, \tag{38}$$

or explicitly

$$\begin{aligned} \mathcal{V}^{(3)} &= \frac{1}{12}\Big((J_1)^2 + (J_2)^2 + J_1.J_2\Big) + \frac{1}{2}[J_1.(j_1+j_3) + J_2.(j_2+j_3)] \\ &+ i\tilde{\psi}_R^j\psi_{R,x}^j - i\tilde{\psi}_L^j\psi_{L,x}^j + im_\psi^j\bar{\psi}^j\psi^j. \end{aligned} \tag{39}$$

The symplectic two-form for this Lagrangian is a non singular matrix, then our algorithm has come to an end. Collecting the canonical part and the symplectic potential $\mathcal{V}^{(3)}$ one has

$$\mathcal{L}[\psi,\overline{\psi}] = \sum_{j=1}^{3}\{i\overline{\psi}^j\gamma^\mu\partial_\mu\psi^j + m_\psi^j\,\overline{\psi}^j\psi^j\} - \sum_{\substack{k,l=1 \\ k\leq l}}^{3}\Big[\bar{a}_{kl}j_k.j_l\Big] + \sum_{l=1}^{3}m^l\nu_l\,j_l^0, \tag{40}$$

where $\nu_3 \equiv \frac{\nu_1+\nu_2}{2}$. We have made the same choice, $k = 1/g$, and the field re-scalings $\psi^j \rightarrow 1/\sqrt{k}\,\psi^j$ as in the last section. This is the same GMT Lagrangian as (17). As a bonus, we get the chemical potentials $\mu_l \equiv m^l \nu_l$ ($\dot{\eta}^{1,2} \rightarrow \nu_{1,2}$) times the charge densities. These terms are related to the charges $Q^l_F = \frac{1}{2\pi}\int_{-\infty}^{+\infty} dx\, j^0_l(t,x)$, and their presence is a consequence of the symplectic method [5].

4.3 The Generalized Sine-Gordon Model (GSG)

One can choose other gauge fixings, instead of (29), to construct the twice-iterated Lagrangian. Let us make the choice

$$\Omega_3 \equiv J^0_1 = 0, \qquad \Omega_4 \equiv J^0_2 = 0, \tag{41}$$

which satisfies the non-gauge invariance condition as can be verified by computing the brackets $\{\Omega_a\ ,\ J^0_b\} = 0;\ a,b = 1,2$. The twice-iterated Lagrangian is obtained by bringing back these constraints into the canonical part of $\mathcal{L}^{(1)}$, then

$$\mathcal{L}^{(2)} = \pi_1\dot{\phi}_1 + \pi_2\dot{\phi}_2 + \dot{\psi}_R\pi^j_R + \dot{\psi}_L\pi^j_L + \dot{\eta}^1\Omega_1 + \dot{\eta}^2\Omega_2 + \dot{\eta}^3\Omega_3 + \dot{\eta}^4\Omega_4 - \mathcal{V}^{(2)}, \tag{42}$$

where the twice-iterated symplectic potential becomes

$$\mathcal{V}^{(2)} = \mathcal{V}^{(1)}\Big|_{\Omega_3=\Omega_4=0}. \tag{43}$$

Considering the set of symplectic variables in the following order

$$\xi^{(2)}_A = (\eta^1, \eta^2, \eta^3, \eta^4, \phi_1, \phi_2, \psi^1_R, \psi^1_L, \psi^2_R, \psi^2_L, \psi^3_R, \psi^3_L, \pi_1, \pi_2, \pi^1_R, \pi^1_L, \pi^2_R, \pi^2_L, \pi^3_R, \pi^3_L) \tag{44}$$

and the components of the canonical one-form

$$a^{(2)}_A = (\Omega_1, \Omega_2, \Omega_3, \Omega_4, \pi_1, \pi_2, -\pi^1_R, -\pi^1_L, -\pi^2_R, -\pi^2_L, -\pi^3_R, -\pi^3_L, 0,0,0,0,0,0,0,0), \tag{45}$$

the (degenerated) 20x20 symplectic matrix is found to be

$$f^{(2)}_{AB}(x,y) = \begin{pmatrix} a_{11} & a_{12} \\ a_{21} & a_{22} \end{pmatrix} \delta(x-y), \tag{46}$$

where
$a_{11} =$

$$\begin{pmatrix}
0 & 0 & 0 & 0 & \partial_x & 0 & im^1\tilde{\psi}^1_R & im^1\tilde{\psi}^1_L & 0 & 0 \\
0 & 0 & 0 & 0 & 0 & \partial_x & 0 & 0 & im^2\tilde{\psi}^2_R & im^2\tilde{\psi}^2_L \\
0 & 0 & 0 & 0 & 0 & 0 & im^1\tilde{\psi}^1_R & im^1\tilde{\psi}^1_L & 0 & 0 \\
0 & 0 & 0 & 0 & 0 & 0 & 0 & 0 & im^2\tilde{\psi}^2_R & im^2\tilde{\psi}^2_L \\
\partial_x & 0 & 0 & 0 & 0 & 0 & 0 & 0 & 0 & 0 \\
0 & \partial_x & 0 & 0 & 0 & 0 & 0 & 0 & 0 & 0 \\
im^1\tilde{\psi}^1_R & 0 & im^1\tilde{\psi}^1_R & 0 & 0 & 0 & 0 & 0 & 0 & 0 \\
im^1\tilde{\psi}^1_L & 0 & im^1\tilde{\psi}^1_L & 0 & 0 & 0 & 0 & 0 & 0 & 0 \\
0 & im^2\tilde{\psi}^2_R & 0 & im^2\tilde{\psi}^2_R & 0 & 0 & 0 & 0 & 0 & 0 \\
0 & im^2\tilde{\psi}^2_L & 0 & im^2\tilde{\psi}^2_L & 0 & 0 & 0 & 0 & 0 & 0
\end{pmatrix},$$

$$a_{12} = \begin{pmatrix}
\frac{im^3}{2}\tilde{\psi}^3_R & \frac{im^3}{2}\tilde{\psi}^3_L & 0 & 0 & m^1\psi^1_R & m^1\psi^1_L & 0 & 0 & \frac{m^3}{2}\psi^3_R & \frac{m^3}{2}\psi^3_L \\
\frac{im^3}{2}\tilde{\psi}^3_R & \frac{im^3}{2}\tilde{\psi}^3_L & 0 & 0 & 0 & 0 & m^2\psi^2_R & m^2\psi^2_L & \frac{m^3}{2}\psi^3_R & \frac{m^3}{2}\psi^3_L \\
\frac{im^3}{2}\tilde{\psi}^3_R & \frac{im^3}{2}\tilde{\psi}^3_L & 0 & 0 & m^1\psi^1_R & m^1\psi^1_L & 0 & 0 & \frac{m^3}{2}\psi^3_R & \frac{m^3}{2}\psi^3_L \\
\frac{im^3}{2}\tilde{\psi}^3_R & \frac{im^3}{2}\tilde{\psi}^3_L & 0 & 0 & 0 & 0 & m^2\psi^2_R & m^2\psi^2_L & \frac{m^3}{2}\psi^3_R & \frac{m^3}{2}\psi^3_L \\
0 & 0 & -1 & 0 & 0 & 0 & 0 & 0 & 0 & 0 \\
0 & 0 & 0 & -1 & 0 & 0 & 0 & 0 & 0 & 0 \\
0 & 0 & 0 & 0 & -1 & 0 & 0 & 0 & 0 & 0 \\
0 & 0 & 0 & 0 & 0 & -1 & 0 & 0 & 0 & 0 \\
0 & 0 & 0 & 0 & 0 & 0 & -1 & 0 & 0 & 0 \\
0 & 0 & 0 & 0 & 0 & 0 & 0 & -1 & 0 & 0
\end{pmatrix},$$

$$a_{21} = \begin{pmatrix}
\frac{im^3}{2}\tilde{\psi}^3_R & \frac{im^3}{2}\tilde{\psi}^3_R & \frac{im^3}{2}\tilde{\psi}^3_R & \frac{im^3}{2}\tilde{\psi}^3_R & 0 & 0 & 0 & 0 & 0 & 0 \\
\frac{im^3}{2}\tilde{\psi}^3_L & \frac{im^3}{2}\tilde{\psi}^3_L & \frac{im^3}{2}\tilde{\psi}^3_L & \frac{im^3}{2}\tilde{\psi}^3_L & 0 & 0 & 0 & 0 & 0 & 0 \\
0 & 0 & 0 & 0 & 1 & 0 & 0 & 0 & 0 & 0 \\
0 & 0 & 0 & 0 & 0 & 1 & 0 & 0 & 0 & 0 \\
m^1\psi^1_R & 0 & m^1\psi^1_R & 0 & 0 & 0 & -1 & 0 & 0 & 0 \\
m^1\psi^1_L & 0 & m^1\psi^1_L & 0 & 0 & 0 & 0 & -1 & 0 & 0 \\
0 & m^2\psi^2_R & 0 & m^2\psi^2_R & 0 & 0 & 0 & 0 & -1 & 0 \\
0 & m^2\psi^2_L & 0 & m^2\psi^2_L & 0 & 0 & 0 & 0 & 0 & -1 \\
\frac{m^3}{2}\psi^3_R & \frac{m^3}{2}\psi^3_R & \frac{m^3}{2}\psi^3_R & \frac{m^3}{2}\psi^3_R & 0 & 0 & 0 & 0 & 0 & 0 \\
\frac{m^3}{2}\psi^3_L & \frac{m^3}{2}\psi^3_L & \frac{m^3}{2}\psi^3_L & \frac{m^3}{2}\psi^3_L & 0 & 0 & 0 & 0 & 0 & 0
\end{pmatrix},$$

$$a_{22} = \begin{pmatrix}
0 & 0 & 0 & \cdots & 0 & -1 & 0 \\
0 & 0 & 0 & \cdots & 0 & 0 & -1 \\
0 & 0 & 0 & \cdots & 0 & 0 & 0 \\
\cdots & \cdots & \cdots & \cdots & \cdots & \cdots & \cdots \\
0 & 0 & 0 & \cdots & 0 & 0 & 0 \\
-1 & 0 & 0 & \cdots & 0 & 0 & 0 \\
0 & -1 & 0 & \cdots & 0 & 0 & 0
\end{pmatrix}.$$

Its zero-modes are

$$\begin{aligned}
\mathbf{v}^{(2)^T}(x) = \Big(&u,\ v,\ \omega,\ \chi,\ 0,\ 0,\ m^1(u+\omega)\psi^1_R,\ m^1(u+\omega)\psi^1_L, \\
&m^2(v+\chi)\psi^2_R,\ m^2(v+\chi)\psi^2_L,\ \frac{m^3}{2}(u+v+\omega+\chi)\psi^3_R, \\
&\frac{m^3}{2}(u+v+\omega+\chi)\psi^3_L,\ u',\ v',\ im^1\tilde{\psi}^1_R(u+\omega), \\
&im^1\tilde{\psi}^1_L(u+\omega),\ im^2\tilde{\psi}^2_R(v+\chi),\ im^2\tilde{\psi}^2_L(v+\chi), \\
&\frac{im^3}{2}\tilde{\psi}^3_R(u+v+\omega+\chi), \frac{im^3}{2}\tilde{\psi}^3_L(u+v+\omega+\chi)\Big),
\end{aligned} \tag{47}$$

where u, v, ω and χ are arbitrary functions. The zero mode condition becomes

$$\int dx \mathbf{v}^{(2)^T}(x)\frac{\delta}{\delta\xi^{(2)}}\int dy' \mathcal{V}^{(2)} = \int dx\, J^1_a\, \partial_x f_a \equiv 0, \quad f_a \equiv (\omega, \chi),\ \ a = 1, 2.$$

Since the functions f_a are arbitrary we end up with the following constraints

$$\Omega_5 \equiv J^1_1 = 0, \quad \Omega_6 \equiv J^1_2 = 0. \tag{48}$$

Notice that by solving the constraints, $\Omega_3 = \Omega_4 = \Omega_5 = \Omega_6 = 0$, eqs. (41) and (48), we may obtain

$$\tilde{\psi}^j_R = \psi^j_R, \qquad \tilde{\psi}^j_L = \psi^j_L. \tag{49}$$

So, at this stage, we have Majorana spinors, the scalars ϕ_1 and ϕ_2, and the auxiliary fields. Next, introduce a third set of Lagrange multipliers into $\mathcal{L}^{(2)}$, then

$$\mathcal{L}^{(3)} = \pi_1\dot{\phi}_1 + \pi_2\dot{\phi}_2 + \dot{\psi}_R\pi_R^j + \dot{\psi}_L\pi_L^j + \dot{\eta}^1\Omega_1 + \dot{\eta}^2\Omega_2 + \dot{\eta}^3\Omega_3 + \dot{\eta}^4\Omega_4 + \dot{\eta}^5\Omega_5 + \dot{\eta}^6\Omega_6 - \mathcal{V}^{(3)}, \quad (50)$$

where

$$\mathcal{V}^{(3)} = \mathcal{V}^{(2)}|_{\Omega_5=\Omega_6=0} \quad (51)$$

or

$$\mathcal{V}^{(3)} = \frac{1}{24}{\phi_{j,x}}^2 + i\psi_R^j\psi_{R,x}^j - i\psi_L^j\psi_{L,x}^j + im_\psi^j\psi_R^j\psi_L^j(e^{-i\phi_j} + e^{i\phi_j}). \quad (52)$$

The new set of symplectic variables is assumed to be ordered as

$\xi_A^{(3)} =$

$$(\eta^1, \eta^2, \eta^3, \eta^4, \eta^5, \eta^6, \phi_1, \phi_2, \psi_R^1, \psi_L^1, \psi_R^2, \psi_L^2, \psi_R^3, \psi_L^3, \pi_1, \pi_2, \pi_R^1, \pi_L^1, \pi_R^2, \pi_L^2, \pi_R^3, \pi_L^3).$$

The components of the canonical one-form are

$a_A^{(3)} =$

$$(\Omega_1, \Omega_2, \Omega_3, \Omega_4, \Omega_5, \Omega_6, \pi_1, \pi_2, -\pi_R^1, -\pi_L^1, -\pi_R^2, -\pi_L^2, -\pi_R^3, -\pi_L^3, 0, 0, 0, 0, 0, 0, 0, 0).$$

After some algebraic manipulations we get the third-iterated 22x22 symplectic two-form

$$f_{AB}^{(3)}(x, y) \quad = \quad \begin{pmatrix} a_{11} & a_{12} \\ a_{21} & a_{22} \end{pmatrix} \delta(x-y). \quad (53)$$

where

$a_{11} =$

$$\begin{pmatrix}
0 & 0 & 0 & 0 & 0 & 0 & \partial_x & 0 & im^1\tilde{\psi}_R^1 & im^1\tilde{\psi}_L^1 & 0 \\
0 & 0 & 0 & 0 & 0 & 0 & 0 & \partial_x & 0 & 0 & im^2\tilde{\psi}_R^2 \\
0 & 0 & 0 & 0 & 0 & 0 & 0 & 0 & im^1\tilde{\psi}_R^1 & im^1\tilde{\psi}_L^1 & 0 \\
0 & 0 & 0 & 0 & 0 & 0 & 0 & 0 & 0 & 0 & im^2\tilde{\psi}_R^2 \\
0 & 0 & 0 & 0 & 0 & 0 & 0 & 0 & -im^1\tilde{\psi}_R^1 & im^1\tilde{\psi}_L^1 & 0 \\
0 & 0 & 0 & 0 & 0 & 0 & 0 & 0 & 0 & 0 & -im^2\tilde{\psi}_R^2 \\
\partial_x & 0 & 0 & 0 & 0 & 0 & 0 & 0 & 0 & 0 & 0 \\
0 & \partial_x & 0 & 0 & 0 & 0 & 0 & 0 & 0 & 0 & 0 \\
im^1\tilde{\psi}_R^1 & 0 & im^1\tilde{\psi}_R^1 & 0 & -im^1\tilde{\psi}_R^1 & 0 & 0 & 0 & 0 & 0 & 0 \\
im^1\tilde{\psi}_L^1 & 0 & im^1\tilde{\psi}_L^1 & 0 & im^1\tilde{\psi}_L^1 & 0 & 0 & 0 & 0 & 0 & 0 \\
0 & im^2\tilde{\psi}_R^2 & 0 & im^2\tilde{\psi}_R^2 & 0 & -im^2\tilde{\psi}_R^2 & 0 & 0 & 0 & 0 & 0
\end{pmatrix},$$

$a_{12} =$

$$\begin{pmatrix}
0 & \frac{im^3}{2}\tilde{\psi}_R^3 & \frac{im^3}{2}\tilde{\psi}_L^3 & 0 & 0 & m^1\psi_R^1 & m^1\psi_L^1 & 0 & 0 & \frac{m^3}{2}\psi_R^3 & \frac{m^3}{2}\psi_L^3 \\
im^2\tilde{\psi}_L^2 & \frac{im^3}{2}\tilde{\psi}_R^3 & \frac{im^3}{2}\tilde{\psi}_L^3 & 0 & 0 & 0 & 0 & m^2\psi_R^2 & m^2\psi_L^2 & \frac{m^3}{2}\psi_R^3 & \frac{m^3}{2}\psi_L^3 \\
0 & \frac{im^3}{2}\tilde{\psi}_R^3 & \frac{im^3}{2}\tilde{\psi}_L^3 & 0 & 0 & m^1\psi_R^1 & m^1\psi_L^1 & 0 & 0 & \frac{m^3}{2}\psi_R^3 & \frac{m^3}{2}\psi_L^3 \\
im^2\tilde{\psi}_L^2 & \frac{im^3}{2}\tilde{\psi}_R^3 & \frac{im^3}{2}\tilde{\psi}_L^3 & 0 & 0 & 0 & 0 & m^2\psi_R^2 & m^2\psi_L^2 & \frac{m^3}{2}\psi_R^3 & \frac{m^3}{2}\psi_L^3 \\
0 & -\frac{im^3}{2}\tilde{\psi}_R^3 & \frac{im^3}{2}\tilde{\psi}_L^3 & 0 & 0 & -m^1\psi_R^1 & m^1\psi_L^1 & 0 & 0 & -\frac{m^2}{2}\psi_R^3 & \frac{m^3}{2}\psi_L^3 \\
im^2\tilde{\psi}_L^2 & -\frac{im^3}{2}\tilde{\psi}_R^3 & \frac{im^3}{2}\tilde{\psi}_L^3 & 0 & 0 & 0 & 0 & -m^2\psi_R^2 & m^2\psi_L^2 & -\frac{m^3}{2}\psi_R^3 & \frac{m^3}{2}\psi_L^3 \\
0 & 0 & 0 & -1 & 0 & 0 & 0 & 0 & 0 & 0 & 0 \\
0 & 0 & 0 & 0 & -1 & 0 & 0 & 0 & 0 & 0 & 0 \\
0 & 0 & 0 & 0 & 0 & -1 & 0 & 0 & 0 & 0 & 0 \\
0 & 0 & 0 & 0 & 0 & 0 & -1 & 0 & 0 & 0 & 0 \\
0 & 0 & 0 & 0 & 0 & 0 & 0 & -1 & 0 & 0 & 0
\end{pmatrix},$$

$a_{21} =$

$$\begin{pmatrix}
0 & im^2\tilde{\psi}_L^2 & 0 & im^2\tilde{\psi}_L^2 & 0 & im^2\tilde{\psi}_L^2 & 0 & 0 & 0 & 0 & 0 \\
\frac{im^3}{2}\tilde{\psi}_R^3 & \frac{im^3}{2}\tilde{\psi}_R^3 & \frac{im^3}{2}\tilde{\psi}_R^3 & \frac{im^3}{2}\tilde{\psi}_R^3 & -\frac{im^3}{2}\tilde{\psi}_R^3 & -\frac{im^3}{2}\tilde{\psi}_R^3 & 0 & 0 & 0 & 0 & 0 \\
\frac{im^3}{2}\tilde{\psi}_L^3 & \frac{im^3}{2}\tilde{\psi}_L^3 & \frac{im^3}{2}\tilde{\psi}_L^3 & \frac{im^3}{2}\tilde{\psi}_L^3 & \frac{im^3}{2}\tilde{\psi}_L^3 & \frac{im^3}{2}\tilde{\psi}_L^3 & 0 & 0 & 0 & 0 & 0 \\
0 & 0 & 0 & 0 & 0 & 0 & 1 & 0 & 0 & 0 & 0 \\
0 & 0 & 0 & 0 & 0 & 0 & 0 & 1 & 0 & 0 & 0 \\
m^1\psi_R^1 & 0 & m^1\psi_R^1 & 0 & -m^1\psi_R^1 & 0 & 0 & 0 & -1 & 0 & 0 \\
m^1\psi_L^1 & 0 & m^1\psi_L^1 & 0 & m^1\psi_L^1 & 0 & 0 & 0 & 0 & -1 & 0 \\
0 & m^2\psi_R^2 & 0 & m^2\psi_R^2 & 0 & -m^2\psi_R^2 & 0 & 0 & 0 & 0 & -1 \\
0 & m^2\psi_L^2 & 0 & m^2\psi_L^2 & 0 & m^2\psi_L^2 & 0 & 0 & 0 & 0 & 0 \\
\frac{m^3}{2}\psi_R^3 & \frac{m^3}{2}\psi_R^3 & \frac{m^3}{2}\psi_R^3 & \frac{m^3}{2}\psi_R^3 & -\frac{m^3}{2}\psi_R^3 & -\frac{m^3}{2}\psi_R^3 & 0 & 0 & 0 & 0 & 0 \\
\frac{m^3}{2}\psi_L^3 & \frac{m^3}{2}\psi_L^3 & \frac{m^3}{2}\psi_L^3 & \frac{m^3}{2}\psi_L^3 & \frac{m^3}{2}\psi_L^3 & \frac{m^3}{2}\psi_L^3 & 0 & 0 & 0 & 0 & 0
\end{pmatrix},$$

$$a_{22} = \begin{pmatrix} 0 & 0 & 0 & 0 & \cdots & 0 & -1 & 0 & 0 \\ 0 & 0 & 0 & 0 & \cdots & 0 & 0 & -1 & 0 \\ 0 & 0 & 0 & 0 & \cdots & 0 & 0 & 0 & -1 \\ 0 & 0 & 0 & 0 & \cdots & 0 & 0 & 0 & 0 \\ \cdots & \cdots & \cdots & \cdots & \cdots & \cdots & \cdots & \cdots & \cdots \\ 0 & 0 & 0 & 0 & \cdots & 0 & 0 & 0 & 0 \\ -1 & 0 & 0 & 0 & \cdots & 0 & 0 & 0 & 0 \\ 0 & -1 & 0 & 0 & \cdots & 0 & 0 & 0 & 0 \\ 0 & 0 & -1 & 0 & \cdots & 0 & 0 & 0 & 0 \end{pmatrix}.$$

It can be checked that this matrix has the zero-modes

$$\begin{aligned} \mathbf{v}^{(3)}(x) &= (u,\, v,\, \omega,\, \chi,\, y,\, z,\, 0,\, 0,\, m^1 a_1^- \psi_R^1,\, m^1 a_1^+ \psi_L^1,\, m^2 a_2^- \psi_R^2,\, m^2 a_2^+ \psi_L^2, \\ & \frac{m^3}{2} a_3^- \psi_R^3,\, \frac{m^3}{2} a_3^+ \psi_L^3,\, u',\, v',\, im^1 a_1^- \tilde{\psi}_R^1,\, im^1 a_1^+ \tilde{\psi}_L^1,\, im^2 a_2^- \tilde{\psi}_R^2,\, im^2 a_2^+ \tilde{\psi}_L^2, \\ & i\frac{m^3}{2} a_3^- \tilde{\psi}_R^3,\, i\frac{m^3}{2} a_3^+ \tilde{\psi}_L^3), \end{aligned} \tag{54}$$

where $a_1^+ \equiv u+\omega+y$, $a_2^+ \equiv v+\chi+z$, $a_3^+ \equiv u+\omega+y+v+\chi+z$, $a_1^- \equiv u+\omega-y$, $a_2^- \equiv v+\chi-z$, $a_3^+ \equiv u+\omega+y+v-\chi-z$, and u, v, ω, χ, y and z are arbitrary functions. The relevant zero-mode condition gives no constraints. Then the action has the following symmetries

$$\delta\xi_A^{(3)} = \mathbf{v}_{\mathbf{A}}^{(3)}(x); \quad A = 1, 2, ...22. \tag{55}$$

These symmetries allow us to fix the bilinears $i\psi_R^j \psi_L^j$ to be constants. By taking $\psi_R^j = -iC_j\overline{\theta}_j$ and $\psi_L^j = \theta_j$, $(j = 1, 2, 3)$ with C_j being real numbers, we find that $i\psi_R^j\psi_L^j$ indeed becomes a constant. Note that θ_j and $\overline{\theta}_j$ are Grassmannian variables, while $\overline{\theta}_j\theta_j$ is an ordinary commuting number.

The two form $f_{AB}^{(3)}(x, y)$, eq. (53), in the subspace $(\phi_1, \phi_2, \pi_{\phi_1}, \pi_{\phi_2})$ defines a canonical symplectic structure modulo canonical transformations. The coordinates ϕ_a and π_{ϕ_a} $(a = 1, 2)$ are not unique. Consider a canonical transformation from (ϕ_a, π_{ϕ_a}) to $(\hat{\phi_a}, \hat{\pi}_{\hat{\phi_a}})$ such that $\phi_a = \frac{\partial F}{\partial \pi_{\phi_a}}$ and $\hat{\phi_a} = \frac{\partial F}{\partial \hat{\pi}_{\hat{\phi}_a}}$. Then, in particular if $\phi_a = \hat{\phi_a}$ one can, in principle, solve for the function F such that a manifestly kinetic term appear in the new Lagrangian.

Then choosing $k = 1/g$ as the overall coupling constant, we are left with

$$\mathcal{L}'' = \sum_{j=1}^{3} \left[\frac{1}{24g}\partial_\mu\phi_j\partial^\mu\phi_j + \frac{M_j}{g}\cos\phi_j\right] + \mu_1\partial_x\phi_1 + \mu_2\partial_x\phi_2, \tag{56}$$

where $M_j = m_\psi^j C_j$. This defines the *generalized sine-Gordon model* (GSG). In addition we get the terms multiplied by chemical potentials μ_1 and μ_2 $(\dot{\eta}^{1,2} \rightarrow -\mu_{1,2})$. These are just the topological charge densities, and are related to the conservation of the number of kinks minus antikinks $Q^a_{\text{topol.}} = \frac{1}{\pi}\int_{-\infty}^{+\infty} dx\ \partial_x\phi_a$.

In the above gauge fixing procedures the possibility of Gribov-like ambiguities deserves a careful analysis. See ref. [5] for a discussion in the $sl(2)$ ATM case. However, in the next section, we provide an indirect evidence of the absence of such ambiguities, at least, for the soliton sector of the model.

Let us comment on the integrability of the GSG model. The usual SG model integrability can be shown by writing the model as a linear problem from which the zero-curvature equation emerges as a compatibility condition. It could be interesting to follow similar approach in the GSG case.

5 The Soliton/Particle Correspondences

The $sl(2)$ ATM theory contains the sine-Gordon (SG) and the massive Thirring (MT) models describing the soliton/particle correspondence of its spectrum [3, 5, 13]. The ATM one-(anti)soliton solution satisfies the remarkable SG and MT classical correspondence in which, apart from the Noether and topological currents equivalence, MT spinor bilinears are related to the exponential of the SG field [24]. The last relationship was exploited in [4] to decouple the $sl(2)$ ATM equations of motion into the SG and MT ones. Here we provide a generalization of that correspondence to the $sl(3)$ ATM case. In fact, consider the relationships

$$\begin{aligned} \frac{\psi_R^1\widetilde{\psi}_L^1}{i} &= -\frac{1}{4\Delta}[(m_\psi^1 p_1 - m_\psi^3 p_4 - m_\psi^2 p_5)\, e^{i(\varphi_2-2\varphi_1)} + m_\psi^2 p_5 e^{3i(\varphi_2-\varphi_1)} \\ &\quad + m_\psi^3 p_4 e^{-3i\varphi_1} - m_\psi^1 p_1] \end{aligned} \tag{57}$$

$$\begin{aligned} \psi_R^2\widetilde{\psi}_L^2 &= -\frac{1}{4\Delta}[(m_\psi^2 p_2 - m_\psi^1 p_5 - m_\psi^3 p_6)\, e^{i(\varphi_1-2\varphi_2)} + m_\psi^1 p_5 e^{3i(\varphi_1-\varphi_2)} \\ &\quad + m_\psi^3 p_6 e^{-3i\varphi_2} - m_\psi^2 p_2] \end{aligned} \tag{58}$$

$$\begin{aligned} \frac{\widetilde{\psi}_R^3\psi_L^3}{i} &= -\frac{1}{4\Delta}[(m_\psi^3 p_3 - m_\psi^1 p_4 + m_\psi^2 p_6)\, e^{i(\varphi_1+\varphi_2)} + m_\psi^1 p_4 e^{3i\varphi_1} \\ &\quad - m_\psi^2 p_6 e^{3i\varphi_2} - m_\psi^3 p_3], \end{aligned} \tag{59}$$

where $\Delta \equiv \bar{a}_{11}\bar{a}_{22}\bar{a}_{33} + 2\bar{a}_{12}\bar{a}_{23}\bar{a}_{13} - \bar{a}_{11}(\bar{a}_{23})^2 - (\bar{a}_{12})^2\bar{a}_{33} - (\bar{a}_{13})^2\bar{a}_{22}$; $p_1 \equiv (\bar{a}_{23})^2 - \bar{a}_{22}\bar{a}_{33}$; $p_2 \equiv (\bar{a}_{13})^2 - \bar{a}_{11}\bar{a}_{33}$; $p_3 \equiv (\bar{a}_{12})^2 - \bar{a}_{11}\bar{a}_{22}$; $p_4 \equiv \bar{a}_{12}\bar{a}_{23} - \bar{a}_{22}\bar{a}_{13}$; $p_5 \equiv \bar{a}_{13}\bar{a}_{23} - \bar{a}_{12}\bar{a}_{33}$; $p_6 \equiv \bar{a}_{11}\bar{a}_{23} - \bar{a}_{12}\bar{a}_{13}$ and the $\bar{a}_{ij}$'s being the current-current coupling constants of the GMT model (17). The relationships (57)-(59) supplied with the conditions (1)-(3) and conveniently substituted into eqs. (109) and (111)-(119) decouple the $sl(3)^{(1)}$ CATM equations into the GSG (56) and GMT (17) equations of motion, respectively.

Moreover, one can show that the GSG (56) M_j parameters and the GMT (17) couplings $\bar{a}_{ij}$ are related by

$$\frac{2\Delta M_1}{g(m_\psi^1)^2} = \bar{a}_{22}(-\frac{m_\psi^3}{m_\psi^1}\bar{a}_{13} + \bar{a}_{33}) + \bar{a}_{23}(-\bar{a}_{23} + \frac{m_\psi^3}{m_\psi^1}\bar{a}_{12}), \tag{60}$$

$$\frac{2\Delta M_2}{g(m_\psi^2)^2} = \bar{a}_{11}(-\frac{m_\psi^3}{m_\psi^2}\bar{a}_{23} + \bar{a}_{33}) + \bar{a}_{13}(-\bar{a}_{13} + \frac{m_\psi^3}{m_\psi^2}\bar{a}_{12}), \tag{61}$$

$$\frac{2\Delta M_3}{g(m_\psi^3)^2} = -\frac{m_\psi^1 m_\psi^2}{(m_\psi^3)^2}(\bar{a}_{12}\bar{a}_{33} - \bar{a}_{13}\bar{a}_{23}) - \bar{a}_{11}\bar{a}_{22} + (\bar{a}_{12})^2. \tag{62}$$

Various limiting cases of the relationships (57)-(59) and (60)-(62) are possible. First, let us consider

$$\bar{a}_{jk} \rightarrow \left\{ \begin{array}{ll} \infty & j = k \neq l, \ \text{(for a given } l) \\ \text{finite} & \text{other cases} \end{array} \right\} \tag{63}$$

then one has

$$\frac{\psi_R^l\widetilde{\psi}_L^l}{i} = \frac{m_\psi^l}{4}\left(e^{-i\phi_l} - 1\right); \quad \psi_R^j\widetilde{\psi}_L^j = 0, \quad j \neq l, \tag{64}$$

for $\bar{a}_{ll} = \delta_l g\,(\delta_{1,2} = 1,\ \delta_3 = -1)$. The three species of one-soliton solutions of the $sl(3)$ ATM theory (4), found in [6] and described in Section 2, satisfy the relationships (64) [4]. Moreover, from eqs. (60)-(62) taking the same limits as in (63) one has

$$M_l = \frac{(m_\psi^l)^2}{2}; \quad M_j = 0, \ j \neq l. \tag{65}$$

Therefore, the relationships (57)-(59) incorporate each $sl(2)$ ATM submodel (particle/soliton) weak/strong coupling phases; i.e., the MT/SG correspondence [4, 5].

Then, the currents equivalence (125), the relationships (57)-(59) and the conditions (1)-(3) satisfied by the one-soliton sector of CATM theory allowed us to establish the correspondence between the GSG and GMT models, thus extending the MT/SG result [24]. It could be interesting to obtain the counterpart of eqs. (57)-(59) for the $N_S \geq 2$ solitons, for example along the lines of [24]. For $N_S = 2$, eq. (125) still holds [6]; and eqs. (1)-(3) are satisfied for the species (α_i, α_i).

Second, consider the limit

$$\bar{a}_{ik} \rightarrow \left\{ \begin{array}{ll} \infty & i = k = j, \;\; (\text{for a chosen } j; \;\; j = 1, 2) \\ \text{finite} & \text{other cases} \end{array} \right\} \tag{66}$$

one gets $M_j = 0$ and

$$4\bar{\Delta}\frac{\psi^l_R \widetilde{\psi}^l_L}{i} = (m^3_\psi \bar{a}_{l3} - m^l_\psi \bar{a}_{33})e^{-i\phi_l} - m^3_\psi \bar{a}_{l3} e^{-3i\varphi_l} + m^l_\psi \bar{a}_{33}, \;\; l \neq j$$

$$4\bar{\Delta}\frac{\psi^3_R \widetilde{\psi}^3_L}{i} = (m^3_\psi \bar{a}_{ll} - m^l_\psi \bar{a}_{l3})e^{-i\phi_3} + m^l_\psi \bar{a}_{l3} e^{-3i\varphi_l} + m^3_\psi \bar{a}_{ll}, \tag{67}$$

$$\psi^j_R \widetilde{\psi}^j_L = 0. \tag{68}$$

where $\bar{\Delta} \equiv 4(\bar{a}_{ll}\bar{a}_{33} - (\bar{a}_{l3})^2)$. The parameters are related by $(m^3_\psi)^2 \bar{a}_{ll} M_l = m^l_\psi (m^3_\psi \bar{a}_{l3} - m^l_\psi \bar{a}_{33}) M_3$. In the case $M_l = M_3 = M$ and redefining the fields as $\phi_l = \sqrt{12g}(A + B)$, $\phi_j = -\sqrt{12g}B$ in the GSG sector, one gets the Lagrangian

$$\mathcal{L}_{BL} = \frac{1}{2}(\partial_\mu A)^2 + \frac{1}{2}(\partial_\mu B)^2 + 2\frac{M}{g}\cos\sqrt{24g}A \, \cos\sqrt{72g}B, \tag{69}$$

which is a particular case of the Bukhvostov-Lipatov model (BL) [25]. It corresponds to a GMT-like theory with two Dirac spinors. The BL model is not classically integrable [26], and some discussions have appeared in the literature about its quantum integrability [27].

Alternatively, if one allows the limit $\bar{a}_{33} \rightarrow \infty$ one gets $\psi^3_R \widetilde{\psi}^3_L = 0$, and additional relations for the ψ^1, ψ^2 spinors and the φ_a scalars. The parameters are related by $\frac{M_1}{(m^1_\psi)^2 \bar{a}_{22}} = \frac{M_2}{(m^2_\psi)^2 \bar{a}_{11}} = -\frac{M_3}{m^1_\psi m^2_\psi \bar{a}_{12}}$. Then we left with two Dirac spinors in the GMT sector and all the terms of the GSG model. The later resembles the 2−cosine model studied in [28] in some sub-manifold of its renormalized parameter space.

6 WZNW Action and the GMT/GSG Duality

In this section we re-derive the previous results in a full Lie algebraic formulation. In section 4.3 an extension of the FJ method was applied to unravel the GSG sector. This method produced the loss of covariance which was restored by means of a complicated canonical transformation, so, here we overcome this problem applying the master Lagrangian method which also provides a final Lagrangian manifestly invariant under the relevant group symmetry of the scalar sector. The off-critical $\hat{sl}(3)$ *affine Toda model coupled to matter fields* (ATM) is defined by the action [8]

$$\begin{aligned} \frac{1}{k} I^{(3)}_{\text{ATM}} &= I_{WZNW}[b] + \int_M d^2x \{ \sum_{m=1}^{2} \Big[< F^-_m \, , \, bF^+_m b^{-1} > \\ &- \frac{1}{2} < E_{-3} \, , \, [W^+_m \, , \, \partial_+ W^+_{3-m}] > + < F^-_m \, , \, \partial_+ W^+_m > \\ &+ \frac{1}{2} < [W^-_m \, , \, \partial_- W^-_{3-m}] \, , \, E_3 > + < \partial_- W^-_m \, , \, F^+_m > \Big] \}, \end{aligned} \tag{70}$$

where

$$I_{WZNW}[b] = \frac{1}{8}\int_M d^2x Tr(\partial_\mu b\partial^\mu b^{-1}) + \frac{1}{12}\int_D d^3x\, \epsilon^{ijk} Tr(b^{-1}\partial_i bb^{-1}\partial_j bb^{-1}\partial_k b), \quad (71)$$

is the Wess-Zumino-Novikov-Witten (WZNW) action.

The W's are some auxiliary fields and the F's and $E_{\pm 3}$ are given in the Appendix. The field b is given by

$$b = e^{i\varphi_1 H_1^0 + i\varphi_2 H_2^0} \quad (72)$$

From the action (70) we get the following matrix equations of motion:

$$\partial_+ W_m^+ = -bF_m^+ b^{-1}, \quad \partial_- W_m^- = -b^{-1}F_m^- b, \quad (73)$$

$$\partial_+(\partial_- bb^{-1}) = \Big[E_{-3}\,,\, bE_l b^{-1}\Big] + \sum_{n=1}^{2}\Big[F_n^-\,,\, bF_n^+ b^{-1}\Big] \quad (74)$$

$$\partial_- F_m^+ = -\Big[E_3\,,\, \partial_- W_{3-m}^-\Big]; \quad \partial_+ F_m^- = \Big[E_{-3}\,,\, \partial_+ W_{3-m}^+\Big]. \quad (75)$$

Taking into account the eqs. (73) and the equations (75) we have the system of equations

$$\partial_+(\partial_- bb^{-1}) = \sum_{n=1}^{2}\Big[F_n^-\,,\, bF_n^+ b^{-1}\Big] \quad (76)$$

$$\partial_- F_m^+ = \Big[E_3\,,\, b^{-1}F_{3-m}^- b\Big], \quad (77)$$

$$\partial_+ F_m^- = -\Big[E_{-3}\,,\, bF_{3-m}^+ b^{-1}\Big], \quad (78)$$

written in terms of the fields b and $F_m^{\pm}$. The system of equations (76)-78) can be obtained from the zero-curvature equation (96) for (97)-(98) with the replacement $B \to b$ provided that the fields satisfy the constraints

$$\Big[F_2^+\,,\, b^{-1}F_1^- b\Big] = 0; \quad \Big[F_2^-\,,\, bF_1^+ b^{-1}\Big] = 0. \quad (79)$$

The equations of motion (109)-(120) are derived from a conformal version of the action (70) [8]. However, by conveniently reducing the conformal model setting to zero the field η, imposing the reality conditions (1) or (2) and the constraints (79), the subset of equations (109) and (111)-(119) can be obtained from the system (76)-(78).

It is an easy task to show that the equations of motion (73)-(75) [or (76)-(78)], as well as the constraints (79), supplied with convenient transformations for the $W_m^{\pm}$ fields, are invariant under the left-right (121)-(124) local symmetries.

6.1 Generalized $\hat{\mathrm{sl}}(3)$ Sine-Gordon Model

In order to uncover the GSG sector of the model (70) we follow the well known technique to perform duality mappings using the master action approach [17]. Following the approach one can remove the derivatives over $x_{\pm}$ in eqs. (75) and write them as

$$F_m^+ = -\Big[E_3\,,\, W_{3-m}^-\Big] - f_m^+(x_+), \quad F_m^- = \Big[E_{-3}\,,\, W_{3-m}^+\Big] + f_m^-(x_-), \quad (80)$$

with $f_m^{\pm}(x_{\pm})$ being analytic functions. Making use of eqs. (80) and (73) one can write the master action (70) as

$$\frac{1}{k}I_{ATM} = I_{WZNW}(b) + \int_M \{-\frac{1}{2}\sum_{m=1}^{2}\Big[2 < f_m^- bf_m^+ b^{-1} > + \\ < [E_{-3}\,,\, W_{3-m}^+]bf_m^+ b^{-1} > + < [E_3\,,\, W_{3-m}^-]b^{-1}f_m^- b >\Big]\} \quad (81)$$

Solving the auxiliary fields $f^{\pm}_{m}$ by means of their Euler-Lagrange equations of motion and replacing into the action (81) one gets

$$\frac{1}{k}I_{ATM} = I_{WZNW}(b) + \frac{1}{4}\int_M \{\sum_{m=1}^{2} < [E_3\,,\,W^{-}_{3-m}]b^{-1}[E_{-3}\,,\,W^{+}_{3-m}]b >\}. \qquad (82)$$

In order to obtain the GSG sector we resort to a convenient gauge fixing procedure. In fact, the action (82) is still invariant under the local symmetries (121)-(124). Therefore, one can make the transformation: $b \rightarrow b'$ and $[E_{\mp 3}\,,\,W^{\pm}_{3-m}] \rightarrow \Lambda^{\mp}_{3-m}$, with $\Lambda^{\pm}_{3-m}$ being constant elements in $\hat{\mathcal{G}}_{\pm(3-m)}$ with similar structure to the $F^{\pm}_{3-m}$. Then the gauge fixed version of (82) becomes

$$\frac{1}{k}I_{ATM} = I_{WZNW}(b) + \int_M \Big[\sum_{m=1}^{2} < \Lambda^{-}_{m} b \Lambda^{+}_{m} b^{-1} > \Big]. \qquad (83)$$

The equation of motion obtained from (83) becomes

$$\partial_{+}(\partial_{-}bb^{-1}) = \sum_{n=1}^{2}\Big[\Lambda^{-}_{n}\,,\,b\Lambda^{+}_{n}b^{-1}\Big]. \qquad (84)$$

The eq. (84) defines the *generalized sine-Gordon model (GSG)*. The parameters $\Lambda^{\pm}_{m}$ have dimension of mass and $k = \kappa/2\pi$, $(\kappa \in \mathbb{Z})$.

Regarding the GMT sector, one can perform a symplectic method reduction starting from the action (70) and taking into account the currents equivalence (125) in matrix form. In dealing with the GMT sector the FJ method turns out to be more appropriate, in particular the Darboux's transformation, which is inherent to this approach, simultaneously diagonalize the spinor canonical one-forms, decouple these fields from the Toda fields, and in addition maintains the algebraic structures of the spinor sector for any affine Lie algebra (see ref. [8] for more details). In the next sub-section instead we will relate the model (84) to a generalized massive Thirring model through a decoupling procedure.

6.2 GMT/GSG Mapping

In this sub-section we compare the field equations of both dual models through some mappings between the spinors (Dirac fields) and the scalars (Toda fields) using the Lie algebraic construction. Thus, we provide the MTM/GSG correspondence by decoupling the ATM eqs. of motion (76)-(78) into the eqs. of motion of the GMT (17) [or (40)] and GSG (56) models, respectively. In fact, consider the relationships [8]

$$\sum_{n=1}^{2}\Big[F^{-}_{n}\,,\,bF^{+}_{n}b^{-1}\Big] = \sum_{n=1}^{2}\Big[\Lambda^{-}_{n}\,,\,b\Lambda^{+}_{n}b^{-1}\Big], \qquad (85)$$

$$\Big[E_{-3}\,,\,bF^{+}_{3-m}b^{-1}\Big] = \Big[E_{-3}\,,\,F^{+}_{3-m}\Big] + \frac{1}{16}\Big[\sum_{n} g_{mn}\Big[F^{+}_{3-n}\,,\,W^{-}_{3-n}\Big]\,,\,F^{-}_{m}\Big], \qquad (86)$$

$$\Big[E_{3}\,,\,b^{-1}F^{-}_{3-m}b\Big] = \Big[E_{3}\,,\,F^{-}_{3-m}\Big] - \frac{1}{16}\Big[\sum_{n} g_{mn}\Big[F^{-}_{3-n}\,,\,W^{+}_{3-n}\Big]\,,\,F^{+}_{m}\Big], \qquad (87)$$

$$F^{\pm}_{m} = \mp[E_{\pm 3}\,,\,W^{\mp}_{3-m}], \qquad (88)$$

$$\Big[F^{\pm}_{2}\,,\,b^{\mp}F^{\mp}_{1}b^{\pm}\Big] = 0, \qquad (89)$$

where the eqs. (89) are the constraints (79) written in a compact form. The $\Lambda^{\pm}_{n}$'s are some constant generators related to the M_j parameters in (56) and assumed to take the same algebraic structure as the corresponding $F^{\pm}_{n}$. The g_{mn} are some coupling constant parameters related to the $\bar{a}_{kl}$'s in (17).

The relationships (86)-(88) when conveniently substituted into (77)-(78) provide

$$\begin{aligned}
\partial_{+}F^{-}_{m} = [E_{-3}\,,\, \partial_{+}W^{+}_{3-m}] &= [E_{-3}\,,\, [E_{3}\,, W^{-}_{m}]] + \\
&\quad \frac{1}{16}[[E_{-3}\,,\, W^{+}_{3-m}]\,, \sum_{n=1}^{2} g_{mn}[[E_{3}\,,\, W^{-}_{n}]\,,\, W^{-}_{3-n}]] \qquad (90)\\
-\partial_{-}F^{+}_{m} = [E_{3}\,,\, \partial_{-}W^{-}_{3-m}] &= -[E_{3}\,,\, [E_{-3}\,, W^{+}_{m}]] - \\
&\quad \frac{1}{16}[[E_{3}\,,\, W^{-}_{3-m}]\,, \sum_{n=1}^{2} g_{mn}[[E_{-3}\,,\, W^{+}_{n}]\,,\, W^{+}_{3-n}]]. \qquad (91)
\end{aligned}$$

This system of equations is the matrix form of the GMT model (17) and can be derived from the action

$$\begin{aligned}
S_{GMT} = \int d^{2}x \sum_{m=1}^{2}\Big(\frac{1}{2} < [E_{-3}\,,\, W^{+}_{3-m}]\partial_{+}W^{+}_{m} > -\frac{1}{2} < [E_{3}\,,\, W^{-}_{3-m}]\partial_{-}W^{-}_{m}] > - \\
< [E_{-3}\,,\, W^{+}_{m}][E_{3}\,,\, W^{-}_{m}] > \Big) - \frac{1}{2}\sum_{m,n} < g_{mn}(j^{+})_{m}\,(j^{-})_{n} >, \qquad (92)
\end{aligned}$$

where the currents are given by

$$j^{0}_{m} = \frac{1}{8}\Big([[E_{-3}\,,\, W^{+}_{m}]\,,\, W^{+}_{3-m}] - [[E_{3}\,,\, W^{-}_{m}]\,,\, W^{-}_{3-m}]\Big) \qquad (93)$$

$$j^{1}_{m} = \frac{1}{8}\Big([[E_{-3}\,,\, W^{+}_{m}]\,,\, W^{+}_{3-m}] + [[E_{l}\,,\, W^{-}_{m}]\,,\, W^{-}_{3-m}]\Big). \qquad (94)$$

The coupling parameters g_{mn} are symmetric by construction, then the current-current interaction terms can be put in a manifestly covariant form. The GMT model (92) includes the kinetic terms, the mass terms $< [E_{-3}\,,\, W^{+}_{m}][E_{3}\,,\, W^{-}_{m}] >$, and the general covariant current-current interaction terms. The canonical pairs are

$$([E_{-3}\,,\, W^{+}_{3-m}]\,,\, W^{+}_{m}) \quad \text{and} \quad ([E_{3}\,,\, W^{-}_{3-m}]\,,\, W^{-}_{m}),$$

where $W^{\pm}_{m} \equiv \mathrm{Tr}\left(W^{\pm}_{m}T^{(\mp m)}_{\lambda(m)}\right)$; $[E_{-3}\,,\, W^{+}_{3-m}] \equiv \mathrm{Tr}\left([E_{-3}\,,\, W^{+}_{3-m}]\,T^{(\pm m)}_{\lambda(m)}\right)$.

Moreover, (85) when substituted into (76) provides the GSG system (84). The GSG system (84)-(83) resemble the Leznov-Saveliev construction of the usual SG theory. It could be an interesting problem to consider an analog construction for the GSG model.

The Noether and topological currents equivalence must also be considered along with the relationships (85)- (89), so the eq. (125) can be written in the form

$$\epsilon^{\mu\nu}\mathrm{Tr}\Big(E^{0}b^{-1}\partial_{\nu}b\Big) = 4\mathrm{Tr}\Big(E^{0}\sum_{m} j^{\mu}_{m}\Big) \qquad (95)$$

with $E^{0} \equiv \mathbf{m}.H^{\pm} = \frac{1}{6}[(2m^{1}_{\psi} + m^{2}_{\psi})H^{0}_{1} + (2m^{2}_{\psi} + m^{1}_{\psi})H^{0}_{2}]$ and the $U(1)$ currents j^{μ}_{m} in matrix form are given in (93)-(94)

The eq. (95) must be considered as an additional relationship between the fields of the GMT and GSG theories.

Using (86)-(88) and the constraints (89) one can establish a linear system of equations for some spinor bilinears with "coefficients" given by certain exponentials of the Toda fields, and by solving this system of equations one obtains the spinor bilinears of the GMT model (17) related to relevant φ_{a} fields of the GSG model (56) which are presented in (57)-(59). These relationships are analogous to those found in [29] for fermion mass bilinears and bosonic ones for massive (free) non-Abelian fermions which are bosonized to certain symmetric space sine-Gordon models. The spinor bilinears expressed in terms of exponentials of Toda fields obtained in this way when conveniently substituted into (85) will provide us an identity equation for the scalars from which on can get some relationships between the parameters of both theories, the couplings g_{mn}, the fermion mass parameters of the GMT model (90)-(90) and the $\Lambda^{\pm}_{m}$ parameters of the GSG theory(84); i.e. the $\bar{a}_{jk}$'s and the M_j relationships as given in (60)-(62).

7 Generalization to Higher Rank Lie Algebra

The procedures presented so far can directly be extended to the CATM model for the affine Lie algebra $sl(n)^{(1)}$ furnished with the principal gradation. According to the construction of [2], these models have soliton solutions for an off-critical sub-model, possess a $U(1)$ vector current proportional to a topological current, apart from the conformal symmetry they exhibit a $\left(U(1)_R\right)^{n-1} \otimes \left(U(1)_L\right)^{n-1}$ left-right local gauge symmetry, and the equations of motion describe the dynamics of the scalar fields φ_a, η, $\widetilde{\nu}\, (a = 1, ...n-1)$ and the Dirac spinors ψ^{α_j}, $\widetilde{\psi}^{\alpha_j}$, $(j = 1, ...N;\ N \equiv \frac{n}{2}(n-1)$ = number of positive roots α_j of the simple Lie algebra $sl(n))$ with one-(anti)soliton solution associated to the field $\alpha_j.\vec{\varphi}$ ($\vec{\varphi} = \sum_{a=1}^{n-1} \varphi_a \alpha_a$, α_a= simple roots of $sl(n)$) for each pair of Dirac fields (ψ^{α_j}, $\widetilde{\psi}^{\alpha_j}$)[2]. Therefore, it is possible to define the off-critical real Lagrangian $sl(n)$ ATM model for the solitonic sector of the theory. The reality conditions would generalize the eqs. (1)-(3); i.e., the new φ's real and the identifications $\widetilde{\psi}^{\alpha_j} \sim (\psi^{\alpha_j})^*$ (up to $\pm$ signs). To apply the symplectic analysis of $sl(n)$ ATM one must impose $(n-1)$ constraints in the Lagrangian, analogous to (5), due to the above local symmetries. The outcome will be a parent Lagrangian of a generalized massive Thirring model (GMT) with N Dirac fields and a generalized sine-Gordon model (GSG) with $(n-1)$ fields. The decoupling of the Toda fields and Dirac fields in the equations of motion of $sl(n)^{(1)}$ CATM, analogous to (109) and (111)-(119), could be performed by an extension of the relationships (57)-(59) and (1)-(3).

The Lagrangian formulation of the $sl(n)^{(1)}$ CATM model and its related off-critical $\hat{sl}(n)$ ATM, as well as the weak-strong duality formulation in matrix form have been presented in [8] thus providing the extension of the results in section 6.

8 Outlook

In the context of the FJ, symplectic, and master Lagrangian approaches we have shown that the $\hat{sl}(3)$ ATM (4) theory is a parent Lagrangian [17] from which both the GMT (17) and the GSG (56) models are derivable. From (17) and (56), it is also clear the duality exchange of the couplings: $g \rightarrow 1/g$. The various soliton/particle species correspondences were uncovered. The soliton sector satifies the $U(1)$ vector and topological currents equivalence (125) and decouples the equations of motion into both dual sectors, through the relationships (57)-(59) (supplied with (1)-(3)). The relationships (57)-(59) contain each $sl(2)$ ATM submodel soliton solution. In connection to these points, recently a parent Lagrangian method was used to give a generalization of the dual theories concept for non p-form fields [30]. In [30], the parent Lagrangian contained both types of fields, from which each dual theory was obtained by eliminating the other fields through the equations of motion.

On the other hand, in nonabelian bosonization of massless fermions [31], the fermion bilinears are identified with bosonic operators. Whereas, in abelian bosonization [32] there exists an identification between the massive fermion operator or Mandelstam operator (charge nonzero sector) and a non-perturbative bosonic soliton operator [33]. Recently, it has been shown that symmetric space sine-Gordon models bosonize the massive nonabelian (free) fermions providing the relationships between the fermions and the relevant solitons of the bosonic model [29]. The ATM model allow us to stablish these type of relationships for interacting massive spinors in the spirit of particle/soliton correspondence. In this context, the generalized Mandelstam soliton operators for the GMT model were recently constructed [19] and the fermion-boson mapping established through a set of generalized bosonization rules in a quotient positive definite Hilbert space of states. In addition, the above approach to the GMT/GSG duality may be useful to construct the conserved currents and the algebra of their associated charges in the context of the CATM $\rightarrow$ ATM reduction. These currents in the MT/SG case were constructed treating each model as a perturbation on a conformal field theory (see [34] and references therein).

The S-duality must be seen at the quantum level (the 2D analogue of the Montonen-Olive conjecture). Regarding this issue the present work provides the algebraic recipe to construct the bosonic dual of the GMT theory (17) as has been recently performed in [19] by showing that the GSG model (56) structure remains almost the same in the full quantum regime.

Finally, using the explicit expressions for the effective actions (83) and (92) corresponding to the GSG and GMT models, respectively, one can study the soliton properties, such as the time delays, in each dual model. In addition, one can obtain the energy-momentum tensors, and hence compute in a direct way the masses of the solitons which were calculated in [2] as the result of the spontaneous breakdown of the conformal symmetry.

Moreover, two dimensional models with four-fermion interactions have played an important role in the understanding of QCD (see, e.g. [35] and references therein). Besides, the GMT model contains explicit mass terms: most integrable models such as the Gross-Neveu, $SU(2)$ and $U(1)$ Thirring models rather all present spontaneous mass generation, the exception being the massive Thirring model. A GMT submodel with $a_{ii} = 0$, $a_{ij} = 1\,(i > j)$ and equal m^j_ψ's, defines the so-called extended Bukhvostov-Lipatov model (BL) and has recently been studied by means of a bosonization technique [36]. Finally, BL type models were applied to $N-$body problems in nuclear physics [37].

Acknowledgements

The author thanks the Mathematics Department of UFMT and Prof. M. da Costa Araújo for hospitality. This work is supported by a CNPq-FAPEMAT grant.

A The $\mathrm{sl}(3)^{(1)}$ CATM Model

We summarize the construction and some properties of the CATM model relevant to our discussions [5]. Consider the zero curvature condition

$$\partial_+ A_- - \partial_- A_+ + [A_+, A_-] = 0. \tag{96}$$

The potentials take the form

$$A_+ = -BF^+B^{-1}, \quad A_- = -\partial_- BB^{-1} + F^-, \tag{97}$$

with

$$F^+ \;=\; E^3 + F_1^+ + F_2^+, \quad F^- \;=\; E^{-3} + F_1^- + F_2^-, \tag{98}$$

where $E^{\pm 3} \equiv \mathbf{m}.H^{\pm} = \frac{1}{6}[(2m^1_\psi + m^2_\psi)H_1^{\pm 1} + (2m^2_\psi + m^1_\psi)H_2^{\pm 1}]$ and the $F_i^{\pm}$'s and B contain the spinor fields and scalars of the model, respectively

$$F_1^+ \;=\; \sqrt{im^1_\psi}\psi^1_R E^0_{\alpha_1} + \sqrt{im^2_\psi}\psi^2_R E^0_{\alpha_2} + \sqrt{im^3_\psi}\widetilde{\psi}^3_R E^1_{-\alpha_3}, \tag{99}$$

[5]Our notations follow that of [7, 8].

$$F_2^+ = \sqrt{im_\psi^3}\psi_R^3 E_{\alpha_3}^0 + \sqrt{im_\psi^1}\widetilde{\psi}_R^1 E_{-\alpha_1}^1 + \sqrt{im_\psi^2}\widetilde{\psi}_R^2 E_{-\alpha_2}^1, \tag{100}$$

$$F_1^- = \sqrt{im_\psi^3}\psi_L^3 E_{\alpha_3}^{-1} - \sqrt{im_\psi^1}\widetilde{\psi}_L^1 E_{-\alpha_1}^0 - \sqrt{im_\psi^2}\widetilde{\psi}_L^2 E_{-\alpha_2}^0, \tag{101}$$

$$F_2^- = \sqrt{im_\psi^1}\psi_L^1 E_{\alpha_1}^{-1} + \sqrt{im_\psi^2}\psi_L^2 E_{\alpha_2}^{-1} - \sqrt{im_\psi^3}\widetilde{\psi}_L^3 E_{-\alpha_3}^0, \tag{102}$$

$$B = e^{i\varphi_1 H_1^0 + i\varphi_2 H_2^0}\, e^{\widetilde{\nu} C}\, e^{\eta Q_{ppal}} \equiv b\, e^{\widetilde{\nu} C}\, e^{\eta Q_{ppal}}, \tag{103}$$

where $E_{\alpha_i}^n$, H_1^n, H_2^n and C ($i = 1,2,3$; $n = 0, \pm 1$) are some generators of $sl(3)^{(1)}$; Q_{ppal} being the principal gradation operator. The commutation relations for an affine Lie algebra in the Chevalley basis are

$$[H_a^m, H_b^n] = mC\frac{2}{\alpha_a^2}K_{ab}\delta_{m+n,0} \tag{104}$$

$$[H_a^m, E_{\pm\alpha}^n] = \pm K_{\alpha a} E_{\pm\alpha}^{m+n} \tag{105}$$

$$[E_\alpha^m, E_{-\alpha}^n] = \sum_{a=1}^r l_a^\alpha H_a^{m+n} + \frac{2}{\alpha^2} mC\delta_{m+n,0} \tag{106}$$

$$[E_\alpha^m, E_\beta^n] = \varepsilon(\alpha,\beta)E_{\alpha+\beta}^{m+n}; \qquad \text{if } \alpha+\beta \text{ is a root} \tag{107}$$

$$[D, E_\alpha^n] = nE_\alpha^n, \qquad [D, H_a^n] = nH_a^n. \tag{108}$$

where $K_{\alpha a} = 2\alpha.\alpha_a/\alpha_a^2 = n_b^\alpha K_{ba}$, with n_a^α and l_a^α being the integers in the expansions $\alpha = n_a^\alpha \alpha_a$ and $\alpha/\alpha^2 = l_a^\alpha \alpha_a/\alpha_a^2$, and $\varepsilon(\alpha,\beta)$ the relevant structure constants.

Take $K_{11} = K_{22} = 2$ and $K_{12} = K_{21} = -1$ as the Cartan matrix elements of the simple Lie algebra $sl(3)$. Denoting by α_1 and α_2 the simple roots and the highest one by $\psi (= \alpha_1 + \alpha_2)$, one has $l_a^\psi = 1 (a = 1,2)$, and $K_{\psi 1} = K_{\psi 2} = 1$. Take $\varepsilon(\alpha,\beta) = -\varepsilon(-\alpha,-\beta)$, $\varepsilon_{1,2} \equiv \varepsilon(\alpha_1,\alpha_2) = 1$, $\varepsilon_{-1,3} \equiv \varepsilon(-\alpha_1,\psi) = 1$ and $\varepsilon_{-2,3} \equiv \varepsilon(-\alpha_2,\psi) = -1$.

One has $Q_{ppal} \equiv \sum_{a=1}^2 \mathrm{s}_a \lambda_a^v.H + 3D$, where λ_a^v are the fundamental co-weights of $sl(3)$, and the principal gradation vector is $\mathrm{s} = (1,1,1)$ [38].

The zero curvature condition (96) gives the following equations of motion

$$\frac{\partial^2\varphi_a}{4i\,e^\eta} = m_\psi^1[e^{\eta - i\phi_a}\widetilde{\psi}_R^l\psi_L^l + e^{i\phi_a}\widetilde{\psi}_L^l\psi_R^l] + m_\psi^3[e^{-i\phi_3}\widetilde{\psi}_R^3\psi_L^3 + e^{\eta+i\phi_3}\widetilde{\psi}_L^3\psi_R^3];\quad a = 1,2 \tag{109}$$

$$-\frac{\partial^2\widetilde{\nu}}{4} = im_\psi^1 e^{2\eta-\phi_1}\widetilde{\psi}_R^1\psi_L^1 + im_\psi^2 e^{2\eta-\phi_2}\widetilde{\psi}_R^2\psi_L^2 + im_\psi^3 e^{\eta-\phi_3}\widetilde{\psi}_R^3\psi_L^3 + \mathbf{m}^2 e^{3\eta}, \tag{110}$$

$$-2\partial_+\psi_L^1 = m_\psi^1 e^{\eta+i\phi_1}\psi_R^1, \qquad -2\partial_+\psi_L^2 = m_\psi^2 e^{\eta+i\phi_2}\psi_R^2, \tag{111}$$

$$2\partial_-\psi_R^1 = m_\psi^1 e^{2\eta-i\phi_1}\psi_L^1 + 2i\Big(\frac{m_\psi^2 m_\psi^3}{im_\psi^1}\Big)^{1/2} e^\eta(-\psi_R^3\widetilde{\psi}_L^2 e^{i\phi_2} - \widetilde{\psi}_R^2\psi_L^3 e^{-i\phi_3}), \tag{112}$$

$$2\partial_-\psi_R^2 = m_\psi^2 e^{2\eta-i\phi_2}\psi_L^2 + 2i\Big(\frac{m_\psi^1 m_\psi^3}{im_\psi^2}\Big)^{1/2} e^\eta(\psi_R^3\widetilde{\psi}_L^1 e^{i\phi_1} + \widetilde{\psi}_R^1\psi_L^3 e^{-i\phi_3}), \tag{113}$$

$$-2\partial_+\psi_L^3 = m_\psi^3 e^{2\eta+i\phi_3}\psi_R^3 + 2i\Big(\frac{m_\psi^1 m_\psi^2}{im_\psi^3}\Big)^{1/2} e^\eta(-\psi_L^1\psi_R^2 e^{i\phi_2} + \psi_L^2\psi_R^1 e^{i\phi_1}), \tag{114}$$

$$2\partial_-\psi_R^3 = m_\psi^3 e^{\eta-i\phi_3}\psi_L^3, \qquad 2\partial_-\widetilde{\psi}_R^1 = m_\psi^1 e^{\eta+i\phi_1}\widetilde{\psi}_L^1, \tag{115}$$

$$-2\partial_+\widetilde{\psi}_L^1 = m_\psi^1 e^{2\eta-i\phi_1}\widetilde{\psi}_R^1 + 2i\Big(\frac{m_\psi^2 m_\psi^3}{im_\psi^1}\Big)^{1/2} e^\eta(-\psi_L^2\widetilde{\psi}_R^3 e^{-i\phi_3} - \widetilde{\psi}_L^3\psi_R^2 e^{i\phi_2}), \tag{116}$$

$$-2\partial_+\widetilde{\psi}_L^2 = m_\psi^2 e^{2\eta-i\phi_2}\widetilde{\psi}_R^2 + 2i\Big(\frac{m_\psi^1 m_\psi^3}{im_\psi^2}\Big)^{1/2} e^\eta(\psi_L^1\widetilde{\psi}_R^3 e^{-i\phi_3} + \widetilde{\psi}_L^3\psi_R^1 e^{i\phi_1}), \tag{117}$$

$$2\partial_-\widetilde{\psi}_R^2 = m_\psi^2 e^{\eta+i\phi_2}\widetilde{\psi}_L^2, \qquad -2\partial_+\widetilde{\psi}_L^3 = m_\psi^3 e^{\eta-i\phi_3}\widetilde{\psi}_R^3, \tag{118}$$

$$2\partial_-\widetilde{\psi}_R^3 = m_\psi^3 e^{2\eta+i\phi_3}\widetilde{\psi}_L^3 + 2i\Big(\frac{m_\psi^1 m_\psi^2}{im_\psi^3}\Big)^{1/2} e^\eta(\widetilde{\psi}_R^1\widetilde{\psi}_L^2 e^{i\phi_2} - \widetilde{\psi}_R^2\widetilde{\psi}_L^1 e^{i\phi_1}), \tag{119}$$

$$\partial^2 \eta \;=\; 0, \tag{120}$$

where $\phi_1 \equiv 2\varphi_1 - \varphi_2$, $\phi_2 \equiv 2\varphi_2 - \varphi_1$, $\phi_3 \equiv \varphi_1 + \varphi_2$.

Apart from the *conformal invariance* the above equations exhibit the $\Big(U(1)_L\Big)^2 \otimes \Big(U(1)_R\Big)^2$ *left-right local gauge symmetry*

$$\varphi_a \;\rightarrow\; \varphi_a + \theta^a_+(x_+) + \theta^a_-(x_-), \quad a = 1,2 \tag{121}$$

$$\widetilde{\nu} \;\rightarrow\; \widetilde{\nu}\,; \qquad \eta \rightarrow \eta \tag{122}$$

$$\psi^i \;\rightarrow\; e^{i(1+\gamma_5)\Theta^i_+(x_+)+i(1-\gamma_5)\Theta^i_-(x_-)}\,\psi^i, \tag{123}$$

$$\widetilde{\psi}^i \;\rightarrow\; e^{-i(1+\gamma_5)(\Theta^i_+)(x_+)-i(1-\gamma_5)(\Theta^i_-)(x_-)}\,\widetilde{\psi}^i, \quad i = 1,2,3; \tag{124}$$

$$\Theta^1_\pm \equiv \pm\theta^2_\pm \mp 2\theta^1_\pm,\; \Theta^2_\pm \equiv \pm\theta^1_\pm \mp 2\theta^2_\pm,\; \Theta^3_\pm \equiv \Theta^1_\pm + \Theta^2_\pm.$$

One can get global symmetries for $\theta^a_\pm = \mp\theta^a_\mp =$ constants. For a model defined by a Lagrangian these would imply the presence of two vector and two chiral conserved currents. However, it was found only half of such currents [6]. This is a consequence of the lack of a Lagrangian description for the $sl(3)^{(1)}$ CATM in terms of the B and $F^\pm$ fields. However, for the sub-model Lagrangian (70) it is possible to construct the Noether currents for each global symmetry.

The gauge fixing of the conformal symmetry, by setting the field η to a constant, is used to stablish the $U(1)$ vector, $J^\mu = \sum_{j=1}^3 m^j_\psi \bar{\psi}^j \gamma^\mu \psi^j$, and topological currents equivalence [2, 5]. Moreover, it has been shown that the soliton solutions are in the orbit of the solution $\eta = 0$. The remarkable equivalence is

$$\sum_{j=1}^{3} m^j_\psi \bar{\psi}^j \gamma^\mu \psi^j \equiv \epsilon^{\mu\nu}\partial_\nu(m^1_\psi\varphi_1 + m^2_\psi\varphi_2), \quad m^3_\psi = m^1_\psi + m^2_\psi, \quad m^i_\psi > 0. \tag{125}$$

References

[1] E. Abdalla, M.C.B. Abdalla and K.D. Rothe, *Non-perturvative methods in two-dimensional quantum field theory*, 2nd edition (World Scientific, Singapore, 2001).

[2] L.A. Ferreira, J-L. Gervais, J. Sánchez Guillen and M.V. Saveliev, *Nucl. Phys.* **B470** (1996) 236.

[3] H. Blas and L.A. Ferreira, *Nucl. Phys.* **B571** (2000) 607.

[4] H. Blas, *Nucl. Phys.* **B596** (2001) 471; H. Blas, *Proc. of VII Hadron Physics 2000 Workshop*, Caraguatatuba, SP, Brazil, 10-15 Apr, 2000, [hep-th/0005037].

[5] H. Blas and B.M. Pimentel, *Annals Phys.* **282** (2000) 67.

[6] A.G. Bueno, L.A. Ferreira and A.V. Razumov, *Nucl. Phys.* **B626** (2002) 463 .

[7] J. Acosta and H. Blas, *J. Math. Phys.* **43** (2002) 1916.

[8] H. Blas, *JHEP* **0311** (2003) 054.

[9] H. Blas, *Phys. Rev.* **D66** (2002) 127701; see also [hep-th/0005130]

[10] S. Brazovskii, *J. Phys. IV* **10** (2000) 169; also in [cond-mat/0006355]; A.J. Heeger, S. Kivelson, J.R. Schrieffer and W. -P. Wu, *Rev. Mod. Phys.* **60** (1988) 782.

[11] D.G. Barci and L. Moriconi, *Nucl. Phys.* **B438** (1995) 522.

[12] R. Jackiw and C. Rebbi, *Phys. Rev.* **D13** (1976) 3398; J. Goldstone and F. Wilczek, *Phys. Rev. Lett.* **47** (1981) 986; J.A. Mignaco and M.A. Rego Monteiro, *Phys. Rev.* **D31** (1985) 3251;

[13] E. Witten, *Nucl. Phys.* **B145** (1978) 110.

[14] L. Faddeev and R. Jackiw, *Phys. Rev. Lett.* **60** (1988) 1692, R. Jackiw, *Diverse Topics in Theoretical Physics*, 1st edition (Worl Scientific 1995).

[15] C. Wotzasek, *Annals Phys.* **243** (1995) 76,
H. Montani and C. Wotzasek, *Mod. Phys. Lett.* **A8** (1993) 3387,
J. Barcelos-Neto and C. Wotzasek, *Mod. Phys. Lett.* **A7** (1992) 1737; *Int. J. Mod. Phys.* **A7** (1992) 4981.

[16] H. Montani, *Int. J. Mod. Phys.* **A8** (1993) 4319.

[17] S. Deser and R. Jackiw, *Phys. Lett.* **139B** (1984) 371; S. E. Hjelmeland and U. Lindstrom, *"Duality for the Nonspecialist"*, [hep-th/9705122].

[18] M.B. Halpern, *Phys. Rev.* **D12** (1975) 1684.

[19] H. Blas, *"Bosonization and generalized Mandelstam soliton operators"*, [hep-th/0404058].

[20] Z. Zhu and D.G. Caldi, *J. Math. Phys.* **36** (1995) 6862; G. Takacs and G. Watts, *Nucl. Phys.* **B547** (1999) 538; S.P. Khastgir and R. Sasaki, *Prog. Theor. Phys.* **95** (1996) 485.

[21] W. Siegel, *Nucl. Phys.* **B238** (1984) 307; M. Henneaux, *Proc. of Summer School in High-Energy Physics and Cosmology* (ICTP-Trieste, 1988).

[22] A.G. Izergin, P.P Kulish, *Letters in Math. Phys.* **2** (1978) 297; *Theor. Mat. Phys.* **44** (1981) 684.

[23] R. Rajaraman; *Solitons and Instantons*, North-Holland, Amsterdam,1982.

[24] S.J. Orfanidis and R. Wang, *Phys. Lett.* **57B** (1975) 281; S.J. Orfanidis, *Phys. Rev.* **D14** (1976) 472.

[25] A.P. Bukhvostov and L.N. Lipatov, *Nucl. Phys.* **B180** (1981) 116.

[26] M. Ameduri and C.J. Efthimiou, *J. Nonl. Math. Phys.* **5** (1998) 132.

[27] H. Saleur, *J. Physics* **A32** (1999) 207; M. Ameduri, C. J. Efthimiou and B. Gerganov, *Mod. Phys. Lett.* **A14** (1999) 2341.

[28] B. Gerganov, *Nucl. Phys.* **B567** (2000) 391.

[29] Q.-Han Park and H.J. Shin, *Nucl. Phys.* **B506** (1997) 537.

[30] H. Casini, R. Montemayor and L.F. Urrutia, *Phys. Lett.* **507B** (2001) 336.

[31] E. Witten, *Commun. Math. Phys.* **92** (1984) 455.

[32] S. Coleman, *Phys. Rev.* **D11** (1975) 2088.

[33] S. Mandelstam, *Phys. Rev.* **D11** (1975) 3026.

[34] R.K. Kaul and R. Rajaraman, *Int. J. Mod. Phys.* **A8** (1993) 1815.

[35] J.F. Bennett and J.A. Gracey, *Nucl. Phys.* **B563** (1999) 390.

[36] J. Sakamoto and B.R. Poudel, *Prog. Theor. Phys.* **104** (2000) 237.

[37] J. Sakamoto and Y. Heike, *Prog. Theor. Phys.* **100** (1998) 399; L.V. Belvedere and R.L.P.G. Amaral, *Phys. Rev.* **D62** (2000) 065009.

[38] V.G. Kac, *Infinite dimensional Lie algebras*, 3rd edition (Cambridge University Press, Cambridge, 1990).

In: New Developments in Soliton Research
Editor: L.V. Chen pp. 149-182
ISBN 1-59554-561-8

Chapter 6

FALACO SOLITONS - COSMIC STRINGS IN A SWIMMING POOL

R.M. Kiehn
Emeritus, Physics Department, University of Houston,
Houston, TX 77004 USA
URL: http://www.cartan.pair.com

Abstract

A dynamical bifurcation mechanism related to changes of metric signature is used to explain the formation of topological defects experimentally observed and defined in 1986 as Falaco Solitons. The Falaco Solitons are topologically coherent structures created experimentally by a macroscopic rotational dynamics in a continuous media with a discontinuity surface, such as that found in a swimming pool. The topological structure of Falaco Solitons replicates certain features found at all physical scales, from spiral arm galaxies and cosmic strings to submicroscopic hadrons. The easy to replicate experiment indicates that the creation of "stationary" thermodynamic states (or solitons) far from equilibrium can be globally stabilized. The Falaco solitons represent a paradigm for explaining a spin pairing mechanism in the microscopic Fermi surface, the development of dimpled vortex structures in rotating Bose-Einstein Condensates, the confinement problem of sub-microscopic quarks on the end of a string connecting branes, and the quantized needle radiation of as photon in terms of a string connecting two Falaco Soliton dimples located on concentric light cone shells.

Keywords: Falaco Solitons, Hopf Breathers, Cosmic Strings between Galaxies, Global Stability, Rotating Bose-Einstein condensates, Bifurcation between Minkowski and Euclidean domains.

1 Falaco Solitons - What Are They?

1.1 A Topologically Coherent Fluid Defect

During March of 1986, while visiting an old MIT friend in Rio de Janeiro, Brazil, the present author became aware of a significant topological event involving visual solitons that can be replicated experimentally by almost everyone with access to a swimming pool. Study the photo which was taken by David Radabaugh, in the late afternoon, Houston, TX 1986.

Figure 1. Falaco Solitons in a Swimming Pool

The extraordinary photo is an image of 3 pairs of what are now called Falaco Solitons, a few minutes after their creation. Each Falaco Soliton consists of a pair of globally stabilized rotational indentations in the water-air discontinuity surface of the swimming pool. The dimple shape is as if a conical pencil point was pushed into a rubber sheet causing a deformation, but the indentation is dominated by dynamic rotation, not translation. Unseen in the photograph, each pair of contra-rotating dimples are connected by a singular thread in the form of a circular arc extending from the vertex of one dimple to the vertex of the other dimple of the pair. The "thread" can be made visible by injecting drops of dye into the fluid near the rotation axis of one of the dimples. These Solitons are apparently long-lived states of matter far from thermodynamic equilibrium. They will persist for many minutes in a still pool of water, maintaining their topological coherence so as to permit their inclusion into the class of objects called Solitons. The Falaco Solitons are extraordinary, not only due to the fact that they are so easily created in a macroscopic dynamical systems environment, but also because they offer real life, easily observed, evidence for the continuous evolution and creation of topological defects.

The long lifetime, and the topological stability, of the Falaco Solitons in a dissipative fluid media is not only remarkable but also is a matter of applied theoretical interest. The equilibrium discontinuity surface of the fluid in the "uniform" g field is flat, and has both zero mean curvature and zero Gauss curvature. The shape of the observed discontinuity surface defect of a Falaco Soliton dimple indicates that the surface mean curvature is zero, but the Gauss curvature is not zero. In Euclidean spaces, such real surfaces are *minimal* surfaces of negative Gauss curvature. Such surfaces are locally unstable, so it has been presumed that the pair of defect structures that make up the Falaco Soliton must be globally stabilized. It has been conjectured that the connecting string is under tension in order to maintain the shape of the pair of dimpled indentations. This conjecture is justified by the observation that if the singular thread is abruptly "severed" (by experimental chopping motions under the surface of the fluid), the dimpled endcaps disappear in a rapid, non-diffusive, manner.

The dimpled surface pairs of the Falaco Soliton are most easily observed in terms of the dramatic black discs that they create by projection of the solar rays to the bottom of the pool. The optics of this effect will be described below. Careful examination of the photo of Figure 1 will indicate, by accidents of noticeable contrast and reflection, the region of the dimpled surface of circular rotation. The dimples appear as (deformed) artifacts to the left of each black spot, and elevated above the horizontal plane by about 25 degrees (as the photo was taken in late afternoon). Also, notice that the vestiges of caustic spiral arms in the surface structures around each pair of rotation axes can be seen. These surface spiral arms can be visually enhanced by spreading chalk dust on the free surface of the pool. The bulk fluid motion is a local (non-rigid body) rotational motion about the interconnecting circular thread. In the photos of Figure 1 and Figure 2, the depth of each of the actual indentations of the free surface is, at most, of a few millimeters in extent.

A better photo, also taken by D. Radabaugh, but in the year 2004 in a swimming pool in Mazan, France, demonstrates more clearly the dimpled surface defects, and the Snell refraction. The sun is to the left and at an elevation of about 30 degrees.

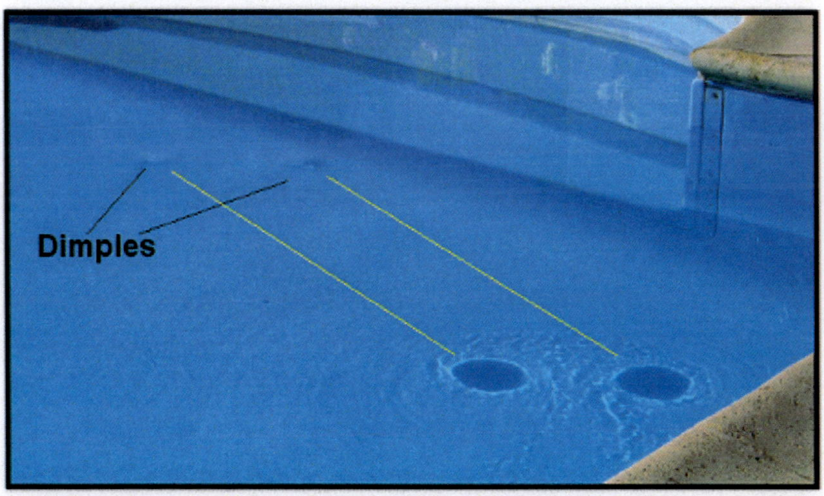

Figure 2. Surface Indentations of a Falaco Soliton

The photo is in effect a single frame of a digital movie that demonstrates the creation and evolutionary motions of the Falaco Solitons. The experimental details of creating the

Falaco Solitons are described below, but the movie explains their creation and dynamics far better than words. The digital movie may be downloaded from [23].

Remark 1. *The bottom line is that it is possible to produce, hydrodynamically, in a viscous fluid with a surface of discontinuity, a long lived topologically coherent structure that consists of a set of macroscopic topological defects. The Falaco Solitons are representative of non-equilibrium long lived structures, or "stationary states", far from equilibrium.*

These observations were first reported at the 1987 Dynamics Days conference in Austin, Texas [14] and subsequently in many other places, mostly in the hydrodynamic literature [15], [16], [20], [22], as well as several APS meetings. More detail is presented in [29].

1.2 Falaco Surface Dimples Are of Zero Mean Curvature

From a mathematical point of view, the Falaco Soliton is interpreted as a connected pair of two dimensional topological defects connected by a one dimensional topological defect or thread. The surface defects of the Falaco Soliton are observed dramatically due the formation of circular black discs on the bottom of the swimming pool. The very dark black discs are emphasized in contrast by a bright ring or halo of focused light surrounding the black disc. All of these visual effects can be explained by means of the unique optics of Snell refraction from a surface of zero mean curvature.

Remark 2. *This explanation of the optics was reached about 30 minutes after I first became aware of the Soliton structures, while standing in the pristine white marble swimming pool of an old MIT roommate, Jose Haraldo Falç ao, under the brilliant Brazilian sunshine in Rio de Janeiro. At MIT, Haraldo was always called Falaco, after he scored 2 goals in a MIT soccer match, and the local newspapers misprinted his name. Hence I dubbed the topological defect structures, Falaco Solitons. Haraldo will get his place in history. I knew that finally I had found a visual, easily reproduced, experiment that could be used to show people the importance and utility of Topological Defects in the physical sciences, and could be used to promote my ideas of Continuous Topological Evolution.*

The observations were highly motivating. The experimental observation of the Falaco Solitons greatly stimulated me to continue research in applied topology, involving topological defects, and the topological evolution of such defects which can be associated with phase changes and thermodynamically irreversible and turbulent phenomena. When colleagues in the physical and engineering sciences would ask "What is a topological defect?" it was possible for me to point to something that they could replicate and understand visually at a macroscopic level.

During the initial few seconds of decay to the metastable soliton state, each large black disk is decorated with spiral arm caustics, remindful of spiral arm galaxies. The spiral arm caustics contract around the large black disk during the stabilization process, and ultimately disappear when the "topological steady" soliton state is achieved. The spiral caustics appear to be swallowed up by the black "hole". It should be noted that if chalk dust is sprinkled on the surface of the pool during the formative stages of the Falaco Soliton, then the topological signature of the familiar Mushroom Spiral pattern is exposed.

Notice that the black spots on the bottom of the pool in the photo are circular and not distorted ellipses, even though the solar elevation is less than 30 degrees. The important experimental fact deduced from the optics of Snell refraction is that each dimpled surface appears to be a surface of zero mean curvature. This conclusion is justified by the fact that the Snell projection to the floor of the pool is almost conformal, preserving the circular appearance of the black disc, independent from the angle of solar incidence. This conformal projection property of preserving the circular shape is a property of normal projection from minimal surfaces of zero mean curvature [24].

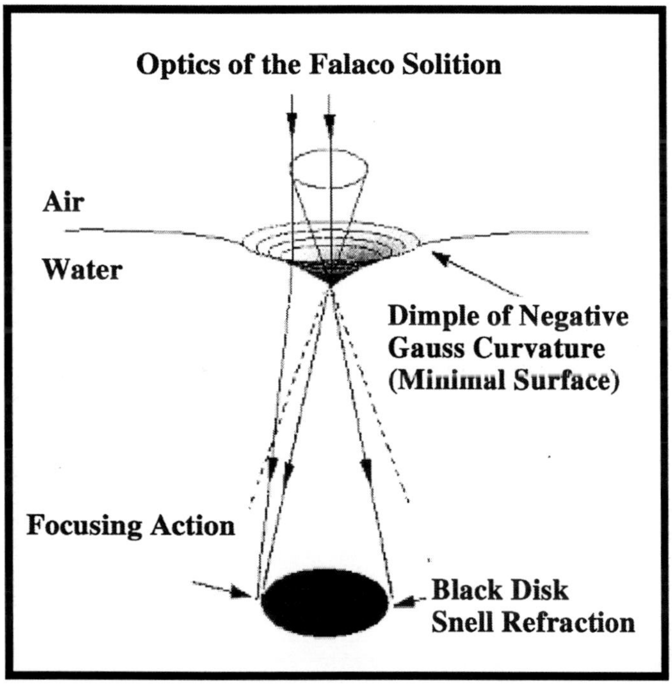

Figure 3. Snell Refraction of a Falaco Soliton surface defect.

As mentioned above, a feature of the Falaco Soliton [14] that is not immediately obvious is that it consists of a pair of two dimensional topological defects, in a surface of fluid discontinuity, which are *connected* by means of a topological singular thread. Dye injection near an axis of rotation during the formative stages of the Falaco Soliton indicates that there is a unseen thread, or 1-dimensional string singularity, in the form of a circular arc that connects the two 2-dimensional surface singularities or dimples. Transverse Torsional waves made visible by dye streaks (caused by dye drops injected near one of the surface rotation axes) can be observed to propagate, back and forth, from one dimple vertex to the other dimple vertex, guided by the "string" singularity. The effect is remindful of the whistler propagation of electrons along the guiding center of the earth's pole to pole magnetic field lines.

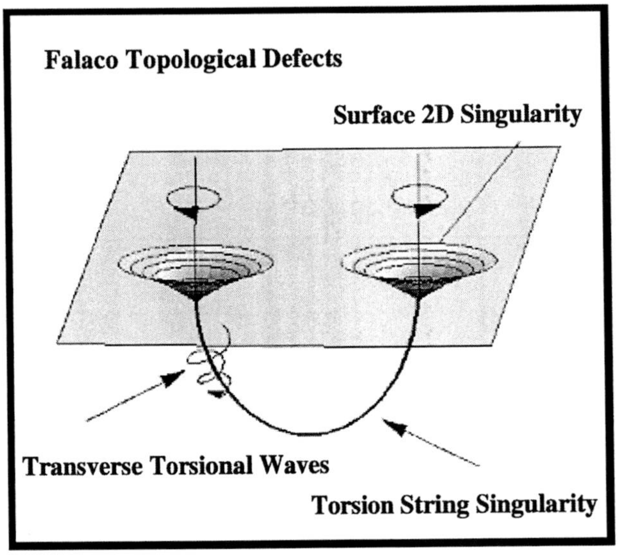

Figure 4. Falaco Topological Defects with connecting thread.

However, as a soliton, the topological system retains its coherence for remarkably long time - more than 15 minutes in a still pool. The long lifetime of the Falaco Soliton is conjectured to be due to this *global stabilization* of the connecting string singularity, even though a real surface of zero mean curvature is locally unstable. The Falaco Soliton starts out from a non-equilibrium thermodynamic state of Pfaff topological dimension 4, which quickly and irreversibly decays to a "topologically stationary" state, still far from equilibrium, but with a long dynamic lifetime [28] [29].

1.3 Falaco Surfaces are Related to Harmonic Vector fields

The long life of the soliton state in the presence of a viscous media indicates that the flow vector field describing the dynamics is probably harmonic. This result is in agreement with the assumption that the fluid can be represented by a Navier-Stokes equation where the viscous dissipation is dominated by affine shear viscosity times the vector Laplacian of the velocity field. If the velocity field is harmonic, the vector Laplacian vanishes, and the shear dissipation term goes to zero - no matter what is the magnitude of the shear viscosity term. Hence a palatable argument is offered in terms of harmonic velocity fields for the existence of the long lifetime of the Falaco Solitons (as well as the production of wakes in fluid dynamics [30]). More over it is known in the theory of minimal surfaces [12] that surfaces of zero mean curvature are generated by harmonic vector fields.

Remark 3. *The bottom line is that the idea of a long lifetime in a dissipative media is to be associated with Harmonic vector fields and surfaces of zero mean curvature.*

Initially it was thought by the present author that the surface configuration, immediately after creation, was in the form of a Rankine vortex (of positive mean curvature, and positive Gauss curvature in a 3D euclidean space), which then decayed into a classic *minimal* surface of zero mean curvature, but negative Gauss curvature. Such an evolutionary process can be found in Langford bifurcations [29] which can be shown to be solutions to the Navier-Stokes equations in a rotating frame of reference. However, such a dynamics seems to require that the connection (the string) between the Falaco pairs has an open throat (like a Wheeler worm hole). Note that for a "stationary" euclidean soap film between two boundary rings (Fig. 5a), the system is stable only if the separation of the boundary rings is less than (approximately) 2.65 times the minimal throat diameter. Experimentally the *stationary non-rotating* soap film between two boundary rings will break apart if the soap film is stretched too far. The single component catenoid (with zero mean curvature and negative Gauss curvature, and with real equal and opposite principle curvatures) will bifurcate into two flat components, one on each ring, and each of zero Gauss curvature as well as zero mean curvature. The process has been demonstrated in fluid flow in a rotating frame, where the zero helicity function of the fluid flow has the appearance of a minimal surface. As the bulk flow increases, the helicity function changes sign, and therefor represents a change in topology from a connected set to a disconnected set. With the change in sign, a torsion bubble (or a torsion burst) appears in the flow pattern [16] [30].

Admittedly, the extended catenoid of Figure 5b (a deformed Wheeler Wormhole with an open throat ?) has the some of the features and appearance of the Falaco Solitons, but the extended singular thread (without an open throat) between vertex singularities does not appear to be replicated.

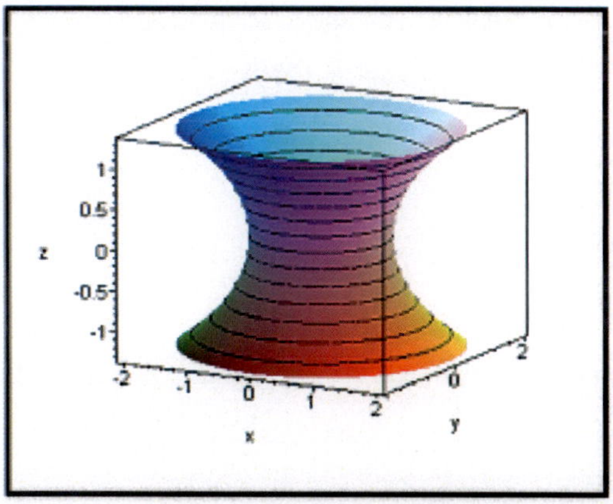

Soap film between rings

Figure 5a.

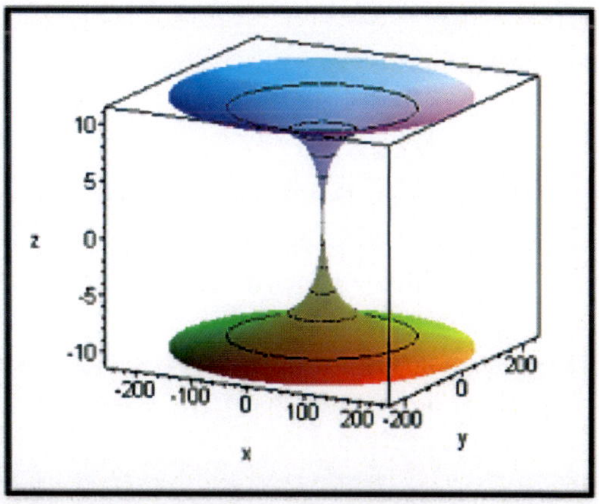

Deformed Wheeler Wormhole

Figure 5b.

It remains difficult to utilize the minimal surface soap film conjecture of a decaying Rankine vortex to support the idea of topological evolution to a structure of two dimpled surfaces of zero mean curvature connected by a 1 dimensional thread. The question arises as how to explain the creation and existence of the Falaco Solitons. The idea that the Falaco Solitons are related to strings connecting branes led to the thought that perhaps the modern advances in topology and string theory could yield a theoretical explanation. According, challenges and requests for help were sent out to many of the string theorists, asking for theoretical help to describe this "real life string connecting branes"; the lack of response indicates that none of the string gurus seemed to think the effort was worthwhile. However, the theoretical work of Dzhunushaliev [4] seems to have many correspondences with the experimental facts of the Falaco Solitons.

In Euclidean space, the real minimal surface defects of zero mean curvature are of negative (or zero) Gauss curvature, and are, therefor, locally unstable. However, stationary non rotating soap films can be stabilized by certain boundary conditions. As mentioned above the experimental equilibrium state of the fluid discontinuity surface is a surface of both zero Gauss curvature and zero Mean curvature (both principle surface curvatures are zero). From the optics of Snell refraction, a Falaco endcap is obviously a surface of zero mean curvature, and if equivalent to a stationary soap film, it should be locally unstable. However, it was conjectured that the local instability could be overcome globally by a string whose tension globally stabilizes the locally unstable endcaps. Could the tension be related to a rotationally induced positive contribution to the otherwise negative Gauss curvature? These conjectures originally were explained (partially) in terms of a bifurcation process and solutions to the Navier-Stokes equations in a rotating frame of reference [29]. A summary of such analysis is presented below.

More recently, it was determined that an alternative, and perhaps better, description can be given in terms of a fluid with a surface discontinuity that has zero mean curvature relative to a Minkowski metric; the Minkowski surface has a Gauss curvature which is positive. A topological birfurcation process from a Rankine vortex to Falaco Solitons would then be such as to change the 3D Euclidean signature into a 3D Minkowski signature. Such surfaces of zero mean curvature embedded in Minkowski space have been called *maximal* surfaces by the differential geometers, and have conical singular points [5] . It is now believed that the Falaco thread is attached to the conical singular points of a pair of such maximal surfaces.

Alternatively, the Euclidean metric can be maintained, and a result similar to the immersion of the 2D surface into Minkowski space can be attributed to the fact that infinitesimal rotations admit Spinor complex isotropic eigen direction fields with non-zero, pure imaginary eigen values. The Gauss curvature of such systems is positive, even though the eigen direction fields are complex Spinors, not vectors in the diffeomorphic sense. A discussion of the Hopf map as applied to this idea will be found below.

1.4 Spinors and Zero Mean Curvature Surfaces

The theory of minimal surfaces (of zero mean curvature) are intertwined with the concept of complex isotropic direction fields, defined as pure Spinors by E.Cartan. The Weirstrass formulas of minimal surface theory [12] consider a holomorphic complex velocity field in 3D, which upon integration leads to conjugate pairs of minimal surfaces defined by the real and imaginary components of the position vector formed by complex integration. The key feature of this holomorphic "velocity" field, so useful to minimal surface theory, is that it is a complex isotropic collection of components, whose Euclidean sums of squares is zero. Such isotropic complex direction fields of zero quadratic form (length) were defined as Spinors by E. Cartan [2].

In addition, E. Cartan demonstrated that infinitesimal rotations are generated by anti-symmetric matrices. It is rather remarkable (and was only fully appreciated by the author in February, 2005) that there is a large class of direction fields (still given the symbol $\rho\mathbf{V}_4$) that do not behave as diffeomorphic vectors. Such direction fields are Spinors and satisfy the equation,

$$\textbf{The Spinor Class: } i(\rho\mathbf{V}_4)dA \neq 0. \tag{1}$$

Spinors are eigen direction fields representing processes, for which the eigenvalues are not zero, but for which the quadratic sums of components squared is zero.

To understand these claims, realize that the Work 1-form is a generalization [28] of the Newtonian concept of Force times Distance:

$$W = i(\rho\mathbf{V}_4)dA = f_m dx^m + Pdt \tag{2}$$

The 2-form dA can be realized as an anti-symmetric matrix of functions. The concept of Work as the 1-form $W = i(\rho\mathbf{V}_4)dA$ focuses attention on the importance of the 2-form, $F = dA$, and its antisymmetric matrix representation, $F \simeq [\mathbb{F}] = -[\mathbb{F}]^{transpose}$. The concept of Work is (in effect) related to the matrix product of $[\mathbb{F}]$ and some vector direction field:

$$W = i(\rho\mathbf{V}_4)dA \simeq [\mathbb{F}] \circ |\rho\mathbf{V}_4\rangle . \tag{3}$$

Suppose $\mathbf{e} = \rho\mathbf{V}_4$ is an eigendirection field with eigenvalue γ, such that

$$i(\rho\mathbf{V}_4)dA \simeq [\mathbb{F}] \circ |\mathbf{e}\rangle = \gamma\,|\mathbf{e}\rangle . \tag{4}$$

Then,

$$i(\rho\mathbf{V})i(\rho\mathbf{V})dA \simeq \langle\mathbf{e}| \circ [\mathbb{F}] \circ |\mathbf{e}\rangle = \gamma\,\langle\mathbf{e}| \circ |\mathbf{e}\rangle . \tag{5}$$

Due to antisymmetry, it follows that

$$i(\rho\mathbf{V})i(\rho\mathbf{V})dA \simeq \langle\mathbf{e}| \circ [\mathbb{F}] \circ |\mathbf{e}\rangle = 0. \tag{6}$$

Hence, for the antisymmetric matrix, $[\mathbb{F}]$, it must be true that

$$\gamma\,\langle\mathbf{e}| \circ |\mathbf{e}\rangle = 0. \tag{7}$$

For division algebras there are two choices: either $\gamma = 0$, or $\langle\mathbf{e}| \circ |\mathbf{e}\rangle = 0$. The implication is that for non zero eigenvalues γ, the quadratic form must vanish:

$$\langle\mathbf{e}| \circ |\mathbf{e}\rangle = (e^1)^2 + (e^2)^2 + ... + (e^n)^2 = 0. \tag{8}$$

Over the real domain, there are no "real vectors" that satisfy this quadratic form, but there are many complex vectors that satisfy the "isotropic" formula. In Euclidean 3 space, the complex integrals of the complex isotropic vectors, when separated into real and imaginary parts, lead to two conjugate 3D "position vectors" that describe immersions of minimal (zero mean curvature) surfaces in 3D.

Remark 4. *The bottom line is that Falaco Solitons can represent non tensorial properties of Spinor analysis, and, as will be developed below, lead to the possibility of surfaces of zero mean curvature, but with positive, not negative, Gauss curvature.*

1.5 Topological Universality Independent from Scales

The reader must remember that the Falaco Soliton is a topological object that can and will appear at all scales, from the microscopic, to the macroscopic, from the sub-submicroscopic world of strings connection branes, to the cosmological level of spiral arm galaxies connected by threads. At the microscopic level, the method offers a view of forming spin pairs that is different from Cooper pairs and could offer insight into Hi-TC Superconductivity. At the level of Cosmology, the concept of Falaco Solitons could lead to explanations of the formation of flat spiral arm galaxies. At the submicroscopic level, the Falaco Solitons mimic quark pairs confined by a string. At the microscopic level, the Falaco Solitons appear as the dimpled vortex structures in rotating Bose-Einstein condensates. They also model the concepts of a Photon as being the singular thread attached to dimples on two expanding light cone shells. At the macroscopic level, similar topological features of the Falaco Solitons can be found in solutions to the Navier-Stokes equations in a rotating frame of reference. Under deformation of the discontinuity surface to a flattened ball, the visual correspondence to hurricane structures between the earth surface and the tropopause is remarkable. In short, as a topological defect, the concept of Falaco Solitons is a universal phenomenon valid at all scales.

1.6 The Experiment

The Falaco Soliton phenomenon is easily reproduced by placing a large circular disc, such as dinner plate, vertically into the swimming pool until the plate is half submerged and it oblate axis resides in the water-air free surface. Then move the plate slowly in the direction of its oblate axis. At the end of the stroke, smoothly extract the plate (with reasonable speed) from the water, imparting kinetic energy and distributed angular momentum to the fluid. The dynamical system undergoes a short period (a few seconds) of stabilization, followed by a longer period (many minutes) of a "topologically stationary" state. It is this topologically stationary state that is defined as the Falaco Soliton. Thermodynamically, the system starts in an initial state of Pfaff topological dimension 4 and decays by continuous topological evolution to a "stationary" state of Pfaff topological dimension 3. According to the theory of non equilibrium thermodynamics [28], the processes during the initial stabilization period are thermodynamically irreversible, but once the Pfaff dimension 3 configuration is reached, the evolutionary processes preserving topological features can be described in a Hamiltonian manner. Both the initial and the "stationary" soliton states are thermodynamic states far from equilibrium.

At first it was thought that the initial deformed surface state could be related to a Rankine vortex structure (which has regions of both positive and negative Gauss curvature). Recall that a Rankine vortex has a core that is equivalent to rigid body rotation. This description of the formative state of stabilization is too naive, for observations indicate that the sharp edge of the plate described above generates instability patterns [30] as it is stroked through the fluid. After the initial injection of energy and angular momentum, the fluid spends a few seconds during a process of stabilization, during which a surface of zero mean curvature is formed transiently, producing the easily visible large black spots formed by Snell refraction. Associated with the evolution to a "stationary" Soliton state, is a visible set of spiral arm caustics on the pool surface around each dimples rotation axis. As the stabilization proceeds, the spiral caustics appear to grow tighter around the black spot, and are almost gone when the Soliton becomes stable.

In a few tries you will become an expert experimentalist at stroking the plate and creating Falaco Solitons. The drifting black spots are easily created and, surprisingly, will persist for many minutes in a still pool. The dimpled depressions are typically of the order of a few millimeters in depth, but the zone of circulation around each rotation axis typically extends over a disc of some 10 to 30 centimeters radius, depending on the plate diameter. The "stationary" configuration, or coherent topological defect structure, has been defined as the Falaco Soliton. For purposes of illustration , the vertical depression has been greatly exaggerated in Figures 3 and 4.

If a thin broom handle or a rod is placed vertically in the pool, and the Falaco soliton pair is directed in its translation motion to intercept the rod symmetrically, as the soliton pair comes within range of the scattering center, or rod, (the range is approximately the separation distance of the two rotation centers) the large black spots at first shimmer and then disappear. Then a short time later, after the soliton has passed beyond the interaction range of the scattering center, the large black spots coherently reappear, mimicking the numerical simulations of soliton coherent scattering. For hydrodynamics, this observation firmly cements the idea that these objects are truly coherent "Soliton" structures. This experiment is the only (known to this author) macroscopic visual experiment that demonstrates these coherence features of soliton scattering.

If the string connecting the two endcaps is sharply "severed", the confined, two dimensional endcap singularities do not diffuse away, but instead disappear almost explosively. The process of "severing" can be accomplished by moving your hand (held under the water approximately above the circular arc or "string" connecting the two dimple vertices) in a karate chop motion. It is this observation that leads to the statement that the Falaco soliton is the macroscopic topological equivalent of the illusive hadron in elementary particle theory. The two 2-dimensional surface defects (the quarks) are bound together by a string of confinement, and cannot be isolated. The dynamics of such a coherent structure is extraordinary, for it is a system that is globally stabilized by the presence of the connecting 1-dimensional string.

For a movie of the process see [23].

2 Bifurcation Process and the Production of Topological Defects

2.1 Lessons from the Bifurcation to Hopf Solitons

2.1.1 Local Stability

Consider a dynamical system that can be encoded (to within a factor, $1/\lambda$) on the variety of independent variables $\{x, y, z, t\}$ in terms of a 1-form of Action:

$$A = \{A_k(x, y, z, t)dx^k - \phi(x, y, z, t)dt\}/\lambda(x, y, z, t). \tag{9}$$

Then construct the Jacobian matrix of the (covariant) coefficient functions:

$$[\mathbb{J}_{jk}(A)] = \left[\partial(A_j/\lambda)/\partial x^k\right] \tag{10}$$

This Jacobian matrix can be interpreted as a projective correlation mapping of "points" (contravariant vectors) into "hyperplanes" (covariant vectors). The correlation mapping is the dual of a collineation mapping, $\left[\mathbb{J}(\mathbf{V}^k)\right]$, which takes points into points. Linear (local) stability occurs at points where the (possibly complex) eigenvalues of the Jacobian matrix are such that the real parts are not positive. The eigenvalues, ξ_k, are determined by solutions to the Cayley-Hamilton characteristic polynomial of the Jacobian matrix, $[\mathbb{J}(A)]$:

$$\Theta(x, y, z, t;\ \xi) = \xi^4 - X_M\xi^3 + Y_G\xi^2 - Z_A\xi + T_K \Rightarrow 0. \tag{11}$$

The Cayley-Hamilton polynomial equation defines a family of implicit functions in the space of variables, $X_M(x, y, z, t)$, $Y_G(x, y, z, t)$, $Z_A(x, y, z, t)$, $T_K(x, y, z, t)$. The functions X_M, Y_G, Z_A, T_K are defined as the similarity invariants of the Jacobian matrix. If the eigenvalues, ξ_k, are distinct, then the similarity invariants are given by the expressions:

$$X_M = \xi_1 + \xi_2 + \xi_3 + \xi_4 = Trace\,[\mathbb{J}_{jk}]\,, \tag{12}$$

$$Y_G = \xi_1\xi_2 + \xi_2\xi_3 + \xi_3\xi_1 + \xi_4\xi_1 + \xi_4\xi_2 + \xi_4\xi_3, \tag{13}$$

$$Z_A = \xi_1\xi_2\xi_3 + \xi_4\xi_1\xi_2 + \xi_4\xi_2\xi_3 + \xi_4\xi_3\xi_1, \tag{14}$$

$$T_K = \xi_1\xi_2\xi_3\xi_4 = \det\,[\mathbb{J}_{jk}]\,. \tag{15}$$

In the differential geometry of 3-dimensional space, $\{x, y, z\}$, when the scaling coefficient is chosen to be the quadratic isotropic Holder norm of index 1 (the Gauss map), then the determinant of the 3x3 Jacobian matrix vanishes, and the resulting similarity invariants become related to the mean curvature and the Gauss curvature of the Shape matrix.

Bifurcation and singularity theory involves the zero sets of the similarity invariants, and the algebraic intersections of the implicit hypersurfaces so generated by these zero sets. Recall that the theory of linear (local) stability requires that the eigenvalues of the Jacobian matrix have real parts which are not greater than zero. For a 4th order polynomial, either all 4 eigenvalues are real; or, two eigenvalues are real, and two eigenvalues are complex conjugate pairs; or there are two distinct complex conjugate pairs. Local stability therefor requires:

Local Stability

$$\text{Odd } X_M \leq 0, \quad \text{Odd } Z_A \leq 0, \tag{16}$$

$$\text{Even } Y_G \geq 0, \quad \text{Even } T_K \geq 0. \tag{17}$$

2.1.2 The Hopf Map

The Hopf map is a rather remarkable projective map from 4 to 3 (real or complex) dimensions that has interesting and useful topological properties related to links and braids and other forms of entanglement. As will be demonstrated, the Hopf map satisfies the criteria of Local Stability, and yet is not an integrable system, and admits irreversible dissipation. The map can be written as $\{x, y, z, s = ct\} \Rightarrow \{x1, x2, x3\}$

$$\textbf{Hopf Map} \quad |\mathbf{H1}\rangle = [x1, x2, x3]^T = [2(xz+ys), 2(xs-yz), (x^2+y^2)-(z^2+s^2)]^T. \tag{18}$$

A remarkable feature of this map is that

$$\langle \mathbf{H}1| \cdot |\mathbf{H}1\rangle = (x1)^2 + (x2)^2 + (x3)^2 = (x^2 + y^2 + z^2 + s^2)^2. \quad (19)$$

Hence a real (and imaginary) 4 dimensional sphere maps to a real 3 dimensional sphere. If the functions $[x1, x2, x3]$ are defined as $[x1/ct, x2/ct, x3/ct]$, then the 4D sphere $(x^2 + y^2 + z^2 + s^2)^2 = 1$, implies that the Hopf map formulas are equivalent to the 4D light cone. The Hopf map can also be represented in terms of complex functions by a map from C2 to R3, as given by the formulas:

$$\mathbf{H}1 = [x1, x2, x3] = [\alpha \cdot \beta^* + \beta \cdot \alpha^*, i(\alpha \cdot \beta^* - \beta \cdot \alpha^*), \alpha \cdot \alpha^* - \beta \cdot \beta^*]. \quad (20)$$

By permuting the formulas it is possible to construct 3 linearly independent Hopf vectors, all of which have same euclidean norm. Note that it is possible to construct complex isotropic spinors by complexifying the Hopf vectors and their permutations:

$$\text{Spinor} \quad |\sigma 1\rangle = |H2\rangle + i\,|H3\rangle\,, \quad \langle \sigma 1| \circ |\sigma 1\rangle = 0\,. \quad (21)$$

It should be expected that there is a connection to surfaces of zero mean curvature and Spinors.

For $\mathbf{H}1$, the 4 independent 1 forms are given by the expressions (where $\Lambda(x, y, z, s)$ is an arbitrary scaling function):

$$d(x1) = 2zd(x) + 2sd(y) + 2xd(z) + 2yd(s) \quad (22)$$

$$d(x2) = 2sd(x) - 2zd(y) - 2yd(z) + 2xd(s) \quad (23)$$

$$d(x3) = 2xd(x) + 2yd(y) - 2zd(z) - sd(s) \quad (24)$$

$$A = \{-yd(x) + xd(y) - sd(z) + zd(s)\}/\Lambda. \quad (25)$$

The formula for the, A, 1-form can be generalized to include constant coefficients of polarization and chirality, to read

$$A_{Hopf} = \{a(-yd(x) + xd(y)) + b(-sd(z) + zd(s))\}/\Lambda. \quad (26)$$

It is some interest to examine the properties of the 1-form, A_{Hopf},defined hereafter as the canonical Hopf 1-form. The Jacobian matrix (for $\Lambda = 1$) becomes

$$JAC_{Hopf} := \begin{bmatrix} 0 & a & 0 & 0 \\ -a & 0 & 0 & 0 \\ 0 & 0 & 0 & -b \\ 0 & 0 & b & 0 \end{bmatrix} \quad (27)$$

with eigen vectors $(e1, e2, e3, e4)$ and eigenvalues $(+ia, -ia, +ib, -ib)$. The similarity invariants are:

$$\text{Odd } X_M = 0, \qquad Z_A = 0. \quad (28)$$

$$\text{Even } Y_G = a^2 + b^2 \geq 0, \quad T_K = a^2 b^2 \geq 0. \quad (29)$$

Hence the canonical Hopf 1-form, A_{Hopf}, is locally stable. It the 1-form is scaled by the factor, $1/\sqrt{(x^2+y^2+z^2+s^2)}$, then the similarity invariants and in all cases represents an imaginary minimal surface. The curvatures are pure imaginary, but the Gauss curvature is positive! For the simple case where $b=0$, the Hopf map describes an minimal surface with imaginary individual curvatures. The classic real minimal surface has a Gauss curvature Y_G which is negative, and for which the individual curvatures are real.

For $\Lambda=1$, it follows that the Hopf 1-form is of Pfaff dimension 4, and has a topological torsion 4-vector proportional to the ray vector from the origin to a point in the space,

$$\mathbf{T}_4 = -2ab[x,y,z,t]\,. \tag{30}$$

Any process that evolves with a component in the direction of $\mathbf{T}_4$ is thermodynamically *irreversible*, as

$$L(\mathbf{T}_4)A = -8ab\;A = Q, \tag{31}$$

$$\text{and } Q\hat{}dQ \neq 0. \tag{32}$$

Consider the Falaco Solitons to be represented by a dynamical system topologically equivalent to an exterior differential system of 1-forms,

$$\omega^k = dx^k - \mathbf{V}^k(x,y,z,t)dt \Rightarrow 0. \tag{33}$$

When all three 1-forms vanish, imposing the existence of a topological limit structure on the base manifold of 4 dimensions, $\{x,y,z,t\}$, the result is equivalent to a 1D solution manifold defined as a kinematic system. The solution manifold to the dynamical system is in effect a parametrization of the parameter t to the space curve $C_{parametric}$ in 4D space, where for kinematic perfection, $[\mathbf{V}^k,1]$ is a tangent vector to the curve $C_{parametric}$. Off the kinematic solution submanifold, the non-zero values for the 1-forms, ω^k, can be interpreted as topological fluctuations from "kinematic perfection".

If "kinematic perfection" is not exact, then the three 1-forms ω^k are not precisely zero, and have a finite triple exterior product that defines a N-1=3 form in the 4 D space. From the theory of exterior differential forms it is the intersection of the zero sets of these three hypersurfaces ω^k that creates an implicit curve $C_{implicit}$ in 4D space.

$$\begin{aligned} C_{implicit} &= \omega^x\hat{}\omega^y\hat{}\omega^z && (34)\\ &= dx\hat{}dy\hat{}dz - \mathbf{V}^x dy\hat{}dz\hat{}dt + \mathbf{V}^y dx\hat{}dz\hat{}dt - \mathbf{V}^z dx\hat{}dy\hat{}dt && (35)\\ &= -i([\mathbf{V},1])\Omega_4. && (36)\end{aligned}$$

The discussion brings to mind the dualism between points (rays) and hypersurfaces (hyperplanes) in projective geometry.

If a ray (a point in a the projective 3 space of 4 dimensions) is specified by the 4 components of a the 4D vector $[\mathbf{V},1]$ multiplied by any non-zero factor, κ, (such that $[\mathbf{V},1] \approx \kappa[\mathbf{V},1]$), then the equation of a dual projective hyperplane is given by the expression $[\mathbf{A},-\phi]$ such that

$$\langle \gamma[\mathbf{A}, -\phi]| \circ |\kappa[\mathbf{V}, 1]\rangle = 0. \tag{37}$$

The principle of projective duality [11] implies that (independent from the factors γ and κ)

$$\phi = \mathbf{A} \circ \mathbf{V}. \tag{38}$$

A particularly easy choice is to assume that (to within a factor)

$$\mathbf{A}_k{=}\mathbf{V}^k, \text{ and } \phi = \mathbf{V} \circ \mathbf{V}, \tag{39}$$

$$A = V_k dx^k - V_k V^k dt. \tag{40}$$

$$V_k(x, y, z, t) \equiv V^k(x, y, z, t), \text{ the 3 functions of a dynamical system.} \tag{41}$$

It should be remembered that not all dynamic features are captured by the similarity invariants of a dynamic system . The antisymmetric features of the dynamics is better encoded in terms of Cartan's magic formula. Cartan's formula expresses the evolution of a 1-form of Action, A, in terms of the Lie differential with respect to a vector field, V, acting on the 1-form that encodes the properties of the physical system. For example, consider the 1-form of Action (the canonical form of a Hopf system) given by the equation

$$A_{Hopf} = \alpha(ydx - xdy) + \beta(tdz - zdt). \tag{42}$$

The Jacobian matrix of this Action 1-form has eigenvalues which are solutions of the characteristic equation,

$$\Theta(x, y, z, t;\, \xi)_{Hopf} = (\xi^2 + \alpha)(\xi^2 + \beta) \Rightarrow 0. \tag{43}$$

The eigenvalues are two conjugate pairs of pure imaginary numbers, $\{\pm i\alpha, \pm i\beta\}$ and are interpreted as "oscillation" frequencies. The similarity invariants are $X_M = 0$, $Y_G = \alpha^2 + \beta^2 > 0$, $Z_A = 0$, $T_K = \alpha^2\beta^2 > 0$. The Hopf eigenvalues have no real parts that are positive, and so the Jacobian matrix is locally stable. The criteria for a double Hopf oscillation frequency requires that the algebraically odd similarity invariants vanish and the algebraically even similarity invariants are positive definite. The stability critical point of the Hopf bifurcation occurs when all similarity invariants vanish. In such a case the oscillation frequencies are zero. This Hopf critical point is NOT necessarily the same as the thermodynamic critical point, as exhibited by a van der Waals gas. The oscillation frequencies have led the Hopf solution to be described as a "breather". The Hopf system is a locally stable system in four dimensions. Each of the pure imaginary frequencies can be associated with a "minimal" hypersurface. .

Suppose that $\beta = 0$. Then the resulting characteristic equation represents a "minimal surface" as $X_M = 0$, but with a Gauss curvature which is positive definite, $Y_G = \alpha^2 > 0$. The curvatures of the implicit surface are imaginary. In differential geometry, where the eigenfunctions can be put into correspondence with curvatures, the Hopf condition, $X_M = 0$, for a single Hopf frequency would be interpreted as "strange" minimal surface (attractor). The surface would be strange for the condition $Y_{G(hopf)} = \alpha^2 > 0$ implies that the Gauss curvature for such a minimal surface is positive. A real minimal surface has curvatures which are real and opposite in sign, such that the Gauss curvature is negative.

As a real minimal surface has eigenvalues with one positive and one negative real number, the criteria for local stability is not satisfied for real minimal surfaces. Yet experience indicates that soap films can occur as "stationary states" when stabilized by certain boundary conditions. The implication is that soap films can be globally stabilized, even though they are locally unstable.

As developed in the next section, the Falaco critical point and the Hopf critical point are the same: all similarity invariants vanish. For the autonomous examples it is possible to find an implicit surface, $Y_{G(hopf)} = Y_{G(falaco)} = 0$, in terms of the variables $\{x, y, z;\ A, B, C...\}$ where $A, B, C...$ are the parameters of the dynamical system.

Recall that the classic (real) minimal surface has real curvatures with a sum equal to zero, but with a Gauss curvature which is negative ($X_M = 0, Y_G < 0$). Such a system is not locally stable, for there exist eigenvalues of the Jacobian matrix with positive real parts. Yet persistent minimal soap films between boundaries can exist under such conditions and are apparently stable macroscopically (globally). This experimental evidence can be interpreted as an example of global stability overcoming local instability.

2.2 The Bifurcation to Falaco Solitons

Similar to and guided by experience with the Hopf bifurcation, the bifurcation that leads to Falaco Solitons must agree with the experimental observation that the endcaps have negative Gauss curvature, and are in rotation. The stability of the Falaco Soliton is global, experimentally, for if the singular thread connecting the vertices is cut, the system decays non-diffusively. Hence the bifurcation to the Falaco Soliton can not imply local stability. This experimental result is related to the theoretical confinement problem in the theory of quarks. To analyze the problem consider the case where the T_K term in the Cayley-Hamilton polynomial vanishes (implying that one eigenvalue of the 4D Jacobian matrix is zero). Experience with the Hopf bifurcation suggests that Falaco Soliton may be related to another form of the characteristic polynomial, where $X_M = 0,\ Z_A = 0,\ Y_G < 0$. This bifurcation is not equivalent to the Hopf bifurcation, but has the same critical point, in the sense that all similarity invariants vanish at the critical point. Similar to the Hopf bifurcation this new bifurcation scheme can be of Pfaff topological dimension 4, which implies that the abstract thermodynamic system generated by the 1-form (which is the projective dual to the dynamical system) is an open, non-equilibrium thermodynamic system. The odd similarity invariants of the 4D Jacobian matrix must vanish. However there are substantial differences between the bifurcation that lead to Hopf solitons (breathers) and Falaco solitons. Experimentally, the Falaco soliton appears to have a projective cusp at the critical point (the vertex of the dimple) and that differs from the Hopf bifurcation which would be expected to have a projective parabola at the critical point.

When $T_K = 0$, the resulting cubic factor of the characteristic polynomial will have 1 real eigenvalue, b, one eigen value equal to zero, and possibly 1 pair of complex conjugate eigenvalues, $(\sigma + i\Omega), (\sigma - i\Omega)$. To be stable globally it is presumed that

Global Stability

$$\text{Odd } X_M = b + 2\sigma \leq 0, \;\; \text{Odd } Z_A = b(\sigma^2 + \Omega^2) \leq 0, \tag{44}$$
$$\text{Even } Y_G = \sigma^2 + \Omega^2 + 2b\sigma \;\; \text{undetermined}, \;\; \text{Even } T_K = 0 \tag{45}$$

If all real coefficients are negative then $Y_G > 0$, and the system is locally stable. Such is the situation for the Hopf bifurcation. However, the Falaco Soliton experimentally requires that $Y_G < 0$.

By choosing $b \leq 0$, in order to satisfy $Z_A \leq 0$, leads to the constraint that $\sigma = -b/2 > 0$, such that the real part of the complex solution is positive, and represents an expansion, not a contraction. Substitution into the formula for Y_G leads to the condition for generation of a Falaco Soliton:

$$Y_{G(falaco)} = \Omega^2 - 3b^2/4 < 0. \tag{46}$$

It is apparent that local stability is lost for the complex eigenvalues of the Jacobian matrix can have positive real parts, $\sigma > 0$. Furthermore it follows that $Y_G < 0$ (leading to negative Gauss curvature) if the square of the rotation speed, Ω, is smaller than the 3/4 of the square of the real (negative) eigen value, b. This result implies that the "forces" of tension overcomes the inertial forces of rotation. In such a situation, a real minimal surface is produced (as visually required by the Falaco soliton). The result is extraordinary for it demonstrates a global stabilization is possible for a system with one contracting direction, and two expanding directions coupled with rotation. The contracting coefficient b (similar to a spring constant) is related to the surface tension in the "string" that connects the two global endcaps of negative Gaussian curvature. The critical point occurs when $\Omega^2 = 3b^2/4$.

It is conjectured that if the coefficient b is in some sense a measure of of a reciprocal length (such that $b \approx 1/R$, a curvature), then there are three interesting formulas comparing angular velocity (orbital period) and length (orbital radius).

$$\text{Falaco}: \;\; \Omega^2 R^2 = \text{constant} \tag{47}$$
$$\text{Kepler}: \;\; \Omega^2 R^3 = \text{constant} \tag{48}$$
$$\text{Planck}: \;\; \Omega^2 R^4 = \text{constant}. \tag{49}$$

The bifurcations to Hopf Solitons suggest oscillations of expansions and contractions of imaginary minimal surfaces (or Soliton concentration breathers) and have been exhibited in the certain chemical reactions. On the other hand, the bifurcations to Falaco Solitons suggest the creation of spiral concentrations, or density waves, on real rotating minimal surfaces.

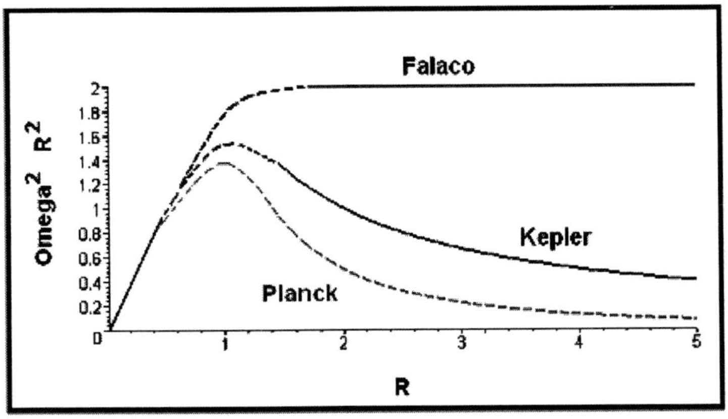

Figure 6. Spiral galaxy mass distributions

The molal density distributions (or order parameters) are complex. The visual bifurcation structures of the Falaco Solitons in the swimming pool would appear to offer an explanation as to the origin of ($\approx$ flat) spiral arm galaxies at a cosmological level, and would suggest that the spiral arm galaxies come in pairs connected by a topological string. Moreover, the kinetic energy of the stars far from the galactic center would not vary as the radius of the "orbit" became very large. This result is counter to the Keplerian result that the kinetic energy of the stars should decrease as 1/R. If is assumed that the density distribution of star mass is more or less constant over the central region of the spiral arm flat disc-like structures, then over this region, the Newtonian gravitation force would lead to a "rigid body" result, $\Omega^2 R^2 = R$. Figure 6. demonstrates the various options:

3 Falaco Solitons in Exact Solutions to the Navier-Stokes Equations.

The idea that multiple parameter Dynamical Systems can produce tertiary bifurcations was studied by Langford [9]. It is remarkable that these tertiary bifurcations can be demonstrated to be solutions of the Navier-Stokes equations in a rotating frame of reference [19]. Langford was interested in how these "normal" forms of dynamical systems could cause bifurcations to Hopf breather-solitons. Herein, it is also of interest to determine how and if these dynamical systems can cause bifurcations to Falaco rotational solitons.

3.1 Minimal Surface Hopf and Falaco Bifurcations

The utility of Maple becomes evident when generalizations of the Langford systems can be studied.

The generalized Langford dynamical system

$$f = A + Bz + Fz^2 + Ez^3 + D(x^2 + y^2) \tag{50}$$

$$g = G + Cz \tag{51}$$

$$dx/dt = \mathbf{V}^x = x(G + Cz) \mp \Omega y \quad (52)$$
$$dy/dt = \mathbf{V}^y = y(G + Cz) \pm \Omega x \quad (53)$$
$$dz/dt = \mathbf{V}^z = A + Bz + Fz^2 + Ez^3 + D(x^2 + y^2) \quad (54)$$

An especially interesting case is given by the system

$$f = A + P\sinh(\alpha z) + D(x^2 + y^2) \quad (55)$$
$$g = G + Cz \quad (56)$$
$$dx/dt = \mathbf{V}^x = x(G + Cz) \mp \Omega y \quad (57)$$
$$dy/dt = \mathbf{V}^y = y(G + Cz) \pm \Omega x \quad (58)$$
$$dz/dt = \mathbf{V}^z = A + P\sinh(\alpha z) + D(x^2 + y^2) \quad (59)$$

Similarity Invariants for the 1-form: $A = V_k dx^k - V^k V_k dt$

$$X_M = 2(G + Cz) + \alpha P\cosh(\alpha z) \quad (60)$$
$$Yg = +\Omega^2 - 2CD(x^2 + y^2) + (G + Cz)^2 + 2(G + Cz)P\alpha\cosh(\alpha z)$$
$$Z_A = (+\Omega^2 + (G + Cz)^2)P\alpha\cosh(\alpha z) - 2CD(G + Cz)(x^2 + y^2)$$
$$T_K = 0 \quad (61)$$

The similarity invariants are polarization invariants relative to the rotation parameter Ω. The criteria for Hopf oscillations requires that $X_M = 0$, and $Z_A = 0$. When these constraints are inserted into the formula for Y_G they yield $Y_{G(hopf)}$. The criteria for oscillations (and breathers) is that $Y_{G(hopf)} > 0$.

$$\text{Hopf Constraint}: Y_{G(hopf)} = 3\Omega^2 - 1/4\alpha^2 P^2(\cosh(\alpha z))^2 > 0 \quad (62)$$
$$\text{Oscillation frequencies: } \omega = \pm\sqrt{-Y_{G(hopf)}} \quad (63)$$

Note that (again) $Y_{G(hopf)}$ is a quadratic form in terms of the rotation parameter. It is therefor easy to identify the tension parameter for the Falaco Soliton by evaluating the Falaco formula

$$Y_{G(falaco)} = \Omega^2 - 3b^2/4. \quad (64)$$
$$\text{Falaco tension } b^2 = (\alpha^2 P^2(\cosh(\alpha z))^2)/3). \quad (65)$$

In this case the tension is again to be associated with a non-linear spring with extensions in the z direction.

$$\text{Helicity} = \mathbf{V} \circ curl\, \mathbf{V}$$
$$H_{bifurcation} = -\{C(x^2 + y^2) + 2(A + P\sinh(\alpha z))\}\Omega.$$

If the process described by the dynamical system is to be reversible in a thermodynamic sense, then the Helicity must vanish. This constraint fixes the value of the rotation frequency Ω in the autonomous system for reversible bifurcations.

The Hopf-Falaco critical point in similarity coordinates can be mapped to an implicit surface in xyz coordinates, eliminating the rotation parameter, Ω.

$$Y_{G(hopf\_critical)} = Y_{G(falaco\_critical)} = -\{3DC(x^2 + y^2) + \alpha^2 P^2(\cosh(\alpha z))^2\} \Rightarrow 0. \tag{66}$$

When the parameters DC have a product which is negative, then the critical surface is the catenoid – *A Minimal Surface*. That is the Hopf critical surface representation of the Gaussian curvature is an implicit surface of given by the equation,

$$(x^2 + y^2) = \{(\alpha^2 P^2)/(3|DC|)\}(\cosh(\alpha z))^2 \tag{67}$$

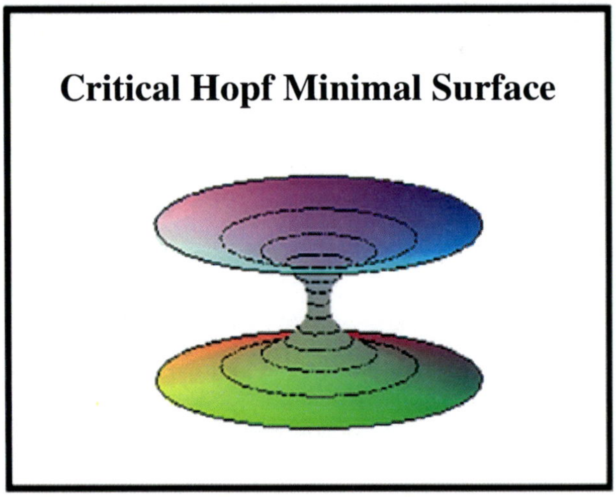

Figure 7. Surface of zero Gauss curvature at the critical point

The catenoid throat diameter is equal to the coefficient $\sqrt{(\alpha^2 P^2)/(3|DC|)}$.

3.2 Maximal Surfaces

Maximal surfaces are 2D surfaces of zero mean curvature that are generated by immersive maps from a two dimension space into a 3 dimensional space with a Lorentz metric [5]. The maximal surface is defined in terms of a space like immersion with positive Gauss curvature and with zero mean curvature. Such maximal surfaces are to be compared to minimal surfaces in a space with a Euclidean metric, but note that minimal surfaces in Euclidean space have negative Gauss curvature. Maximal surfaces can admit isolated, or "conical", singularities, where Minimal surfaces do not. Maximal surfaces can mimic catenoidal and helical surfaces of Euclidean theory, but may exhibit singular subsets of points. It is remarkable (and discussed in the next section) that such maximal surfaces can appear in fluids as propagating long lived topological defects which have been described above as Falaco Solitons.

Consider a 3D space with a Minkowski - Lorentz metric of the form

$$(ds)^2 = (dx)^2 + (dy)^2 - (dz)^2. \tag{68}$$

The immersion

$$R(u, v) = [(\sinh(v)\cos(u), (p\sinh(a)\sin(u), h\,v] \tag{69}$$

generates a surface of zero mean curvature in a space with a Minkowski metric. The coefficient p is related to the handedness of the rotation about the z axis, and h is related to the helicity along the z axis. The surface is of zero mean curvature, but the metric vanishes at the conical singular point: the Gauss curvature becomes infinite. The immersion does not generate a minimal surface in euclidean space. For Other examples of zero mean curvature surfaces in both Euclidean and Minkowski spaces see [10]

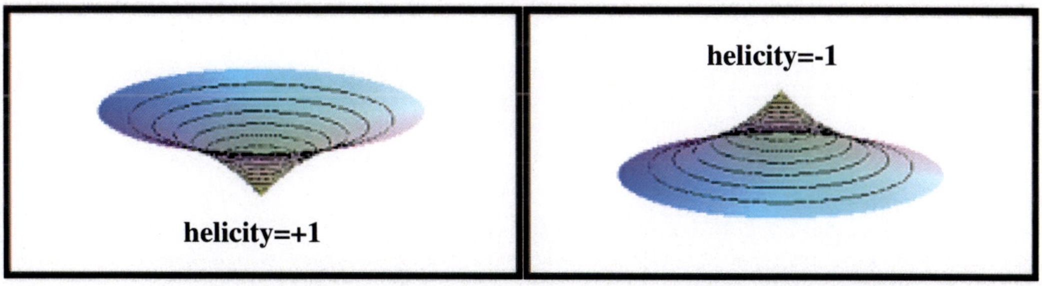

Figures 8a. and 8b. Minkowski surfaces of zero mean curvature

The surface is similar to the hyperbolic minimal surface (Catenoid) in Euclidean geometry, but here, unlike the Euclidean catenoid, the Minkowski catenoid has a singular point. The surface is sensitive to the sign of the directional chirality ($h = \pm 1$), but is not sensitive to the handedness of polarization, p.

3.2.1 Immersions That Do Not Depend upon the 3D Signature

The hyperbolic rotational immersion,

$$R(u, v) = [\cosh(v)\cos(u), p\cosh(v)\sin(u), hv], \tag{70}$$

generates a minimal surface of zero mean curvature both in a space with an Euclidean metric, or in a space with a Minkowski metric. The surface "mimics" a Wheeler wormhole, and the soap film between two rings separated by a diameter. The zero mean curvature surface is also sensitive to the sign of the directional chirality ($h = \pm 1$), but is not sensitive to the handedness of rotational polarization, p.

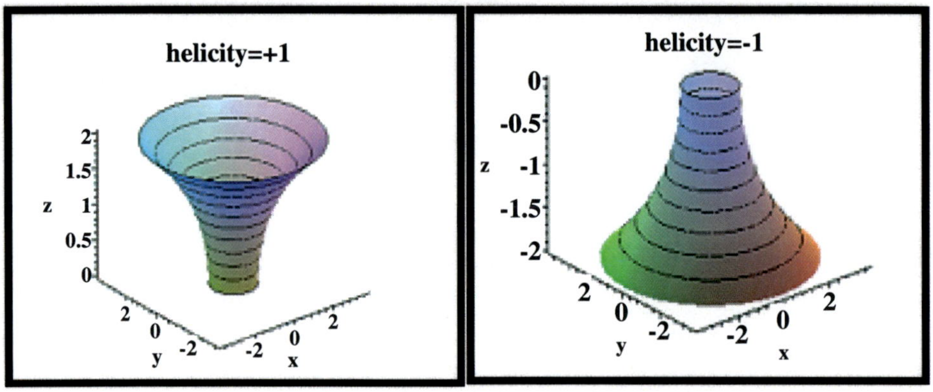

Fig. 9a and 9b. Rotational Surfaces zero mean curvature independent of 3D signature

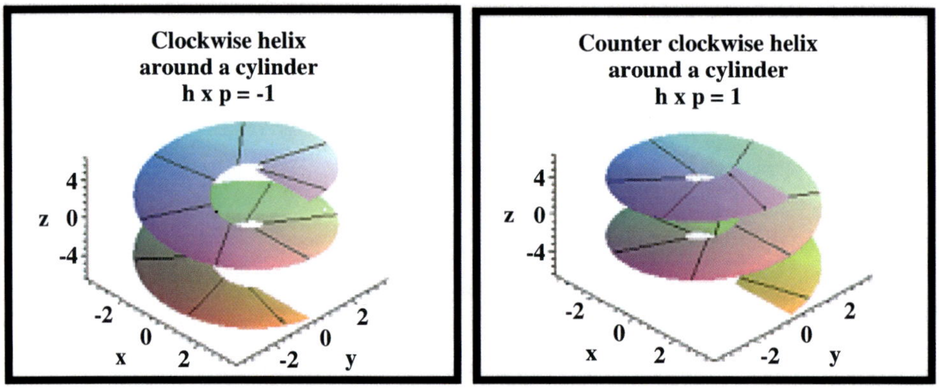

Figure 10a. and 10b. Helical Surfaces of zero mean curvature independent from 3D signature.

On the other hand, the surface generated by the hyperbolic helical immersion

$$R(u, v) = [(\cosh(v)\cos(u), (p\cosh(v)\sin(u), h\, u] \tag{71}$$

also is a surface of zero mean curvature in *both* a Euclidean space or in Minkowski space.

The surfaces are ruled helices rapped around a "hole" of radius unity. The Helical surface is sensitive to the sign of the product of the rotational polarization, p, and the directional chirality, h. The Gauss curvature of the immersion is negative and bounded in Euclidean Space. The Gauss curvature of the immersion is positive and singular for $v = 0$ in the Minkowski space. Both surfaces have zero mean curvature. The principal surface curvatures are real and of opposite in sign for the Euclidean 3 space, and are pure imaginary and of opposite sign in Minkowski 3 space. In both cases the Gauss curvature is real but of different signs.

The zero mean curvature surfaces, with a singular point (as in Figures 8a. and 8b.), can be formed experimentally in a fluid. The experimental evidence is presented below. The idea that 3-dimensional space may or may not be visually Euclidean challenges a dogmatic precept of modern physics, where it is rarely perceived that physical 3D space can be anything but Euclidean. However, as discussed in the following section, the occurrence of long lived rotational structures in the free surface of a water, which have been described as Falaco Solitons, exhibit the features of maximal surfaces in a Lorentz - Minkowski space. The Falaco Solitons are topological defect structures easily replicated in an experimental sense. Optical measurements indicate that the surface defect structures have a zero mean curvature. In addition, the surface defect structures have an apparent conical singularity which is an artifact of the signature of a maximal space-like surface in Minkowski space. Maximal surfaces are generated by immersive maps from a two dimension space into a 3 dimensional space with a Lorentz metric [5]. The maximal surface is defined in terms of a space like immersion with positive Gauss curvature and with zero mean curvature. Such surfaces are related to minimal surfaces in a space with a Euclidean metric, but minimal surfaces in Euclidean space have negative Gauss curvature. Maximal Surfaces can admit isolated, or "conical", singularities, where Minimal surfaces can not. Maximal surfaces can mimic catenoidal and helical surfaces of Euclidean theory, but may exhibit singular subsets of points.

The zero mean curvature surfaces, with a singular point, can be formed experimentally in a fluid. The experimental evidence is given by the existence of the Falaco Solitons. The idea that 3-dimensional space may or may not be Euclidean challenges a dogmatic precept of modern physics, where it is rarely perceived that physical 3D space can be anything but Euclidean. However, as discussed above, the occurrence of long lived rotational structures in the free surface of a water, which have been described as Falaco Solitons, exhibit the features of maximal surfaces in a Lorentz - Minkowski space. The Falaco Solitons are topological defect structures easily replicated in an experimental sense. Optical measurements indicate that the surface defect structures have a zero mean curvature. In addition, the surface defect structures have an apparent conical singularity which is an artifact of the signature of a maximal space-like surface in Minkowski space.

The conjecture is that the Falaco Solitons are topological defects caused by the decay of a dissipative Pfaff dimension 4 domain, with a spacelike Euclidean structure, followed by a topological bifurcation process that changes the space-like Sylvester signature from a 3D Euclidean structure to a space like 3D Minkowski structure.

4 Landau Ginsburg structures, Falaco Solitons and Spiral Arm Structures

The Falaco experiments lead to the idea that such topological defects are available at all scales, and are related to fourth order Landau - Ginsburg structures. Figure 11 is adapted from Tornkvist [26].

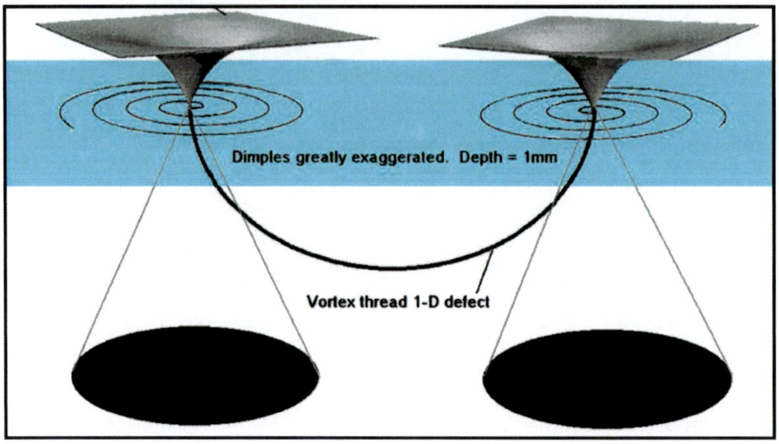

Figure 11. Falaco Solitons related to Landau Ginsburg theory

The Falaco Solitons consist of spiral "vortex defect" structures (analogous to CGL theory) on a two dimensional minimal surface, one at each end of a 1-dimensional "vortex line" or thread (analogous to GPG theory). Remarkably the topological defect surface structure is locally unstable, if the surface is of negative Gauss curvature. Yet the pair of locally unstable 2-D surfaces is *globally* stabilized by the 1-D line defect attached to the "vertex" points of the minimal surfaces. For some specific physical systems it can be demonstrated that period (circulation) integrals of the 1-form of Action potentials, A, lead to the concept of "vortex defect lines". The idea is extendable to "twisted vortex defect lines" in three dimensions. The "twisted vortex defects" become the spiral vortices of a Complex Ginsburg Landau (CGL) theory , while the "untwisted vortex lines" become the defects of Ginzburg-Pitaevskii-Gross (GPG) theory [26].

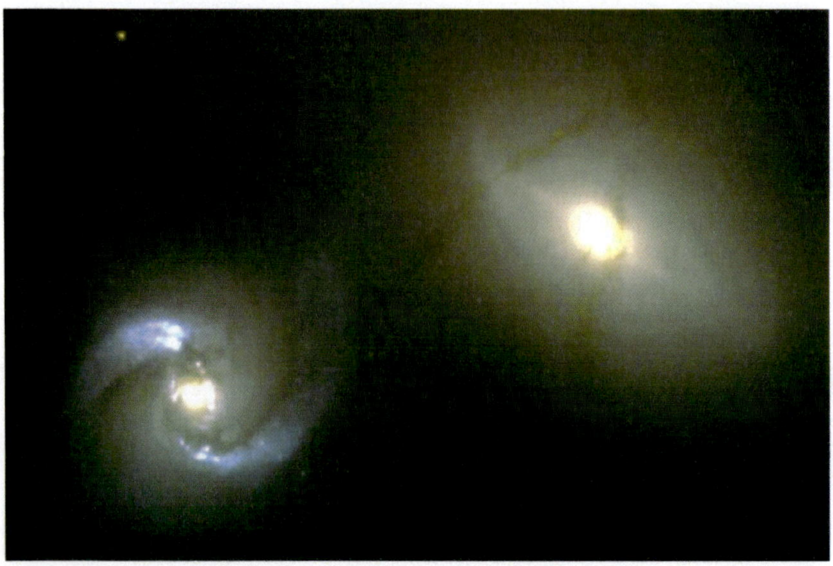

Figure 12. Hubble photo: Cosmic strings linking spiral arm galaxies?

In the macroscopic domain, the experiments visually indicate "almost flat" spiral arm structures during the formative stages of the Falaco solitons. In the cosmological domain, it is suggested that these universal topological defects represent the ubiquitous "almost flat" spiral arm galaxies. Based on the experimental creation of Falaco Solitons in a swimming pool, it has been conjectured that M31 and the Milky Way galaxies could be connected by a topological defect thread [29]. Only recently has photographic evidence appeared suggesting that Galaxies may be connect by strings.

4.1 Wheeler Wormholes and Falaco Strings between Branes

It is extraordinary, but the Falaco Solitons appear to be another form of a zero mean curvature surface structure, either related to macroscopic realizations of the Wheeler wormhole (with a very narrow throat), or to Spinor surfaces generated by complex eigen direction fields of infinitesimal rotations. The Wheeler wormhole structure was presented early on by Wheeler (1955), but was considered to be unattainable in a practical sense. To quote Zeldovich p. 126 [34]

> "The throat or "wormhole" (in a Kruskal metric) as Wheeler calls it, connects regions of the same physical space which are extremely remote from each other. (Zeldovich then gives a sketch that topologically is similar to the Falaco Soliton). Such a topology implies the existence of 'truly geometrodynamic objects' which are unknown to physics. Wheeler suggests that such objects have a bearing on the nature of elementary particles and anti particles and the relationships between them. However, this idea has not yet borne fruit; and there are no macroscopic"geometrodynamic objects" in nature that we know of. Thus we shall not consider such a possibility."

This quotation dates back to the period 1967-1971.

Now the experimental evidence justifies (again) Wheeler's intuition. Both the Wheeler wormhole and the Falaco Soliton are related to surface structures of zero mean curvature. The catenoidal surface of zero mean curvature, and negative Gauss curvature, in a 3D Euclidean space is a Wheeler Wormhole (with an open throat), while the conical surface of zero mean curvature, and positive Gauss curvature, and its conical singular point in a 3D Minkowski space is a part of the rotationally induced Falaco Soliton.

Remark 5. *The bottom line is that the remarkable features of creating a stable surface of zero mean curvature and positive Gauss curvature (the Falaco Soliton) is explained either by assuming that the usual 3D Euclidean Signature is rotationally dependent and can topologically evolve into a 3D Minkowski Signature; or, the Euclidean Signature is preserved, and a macroscopic evolutionary process described by complex Spinor direction fields (which are not the same as diffeomorphic vector fields) must be admitted on thermodynamic grounds.*

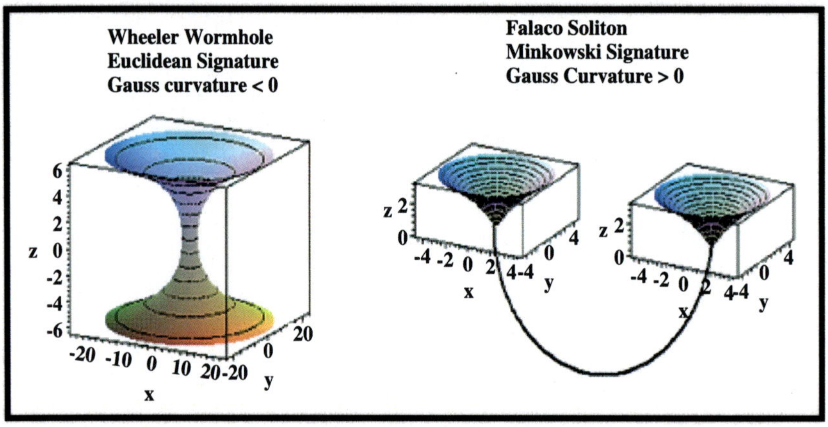

Figure 13. Rotational Surfaces of Zero mean curvature in Euclidean and Minkowski 3 space

The Falaco Soliton endcap dimples (which are presumed to be surfaces of zero mean curvature and positive Gauss curvature) are related to Spinor eigen direction fields associated with antisymmetric matrices representing Symplectic spaces. If the Maximal surfaces appear as deformations in disconnected hypersurfaces of discontinuity, the topological structure has the appearance of "strings connecting branes", a concept touted by the string theorists .

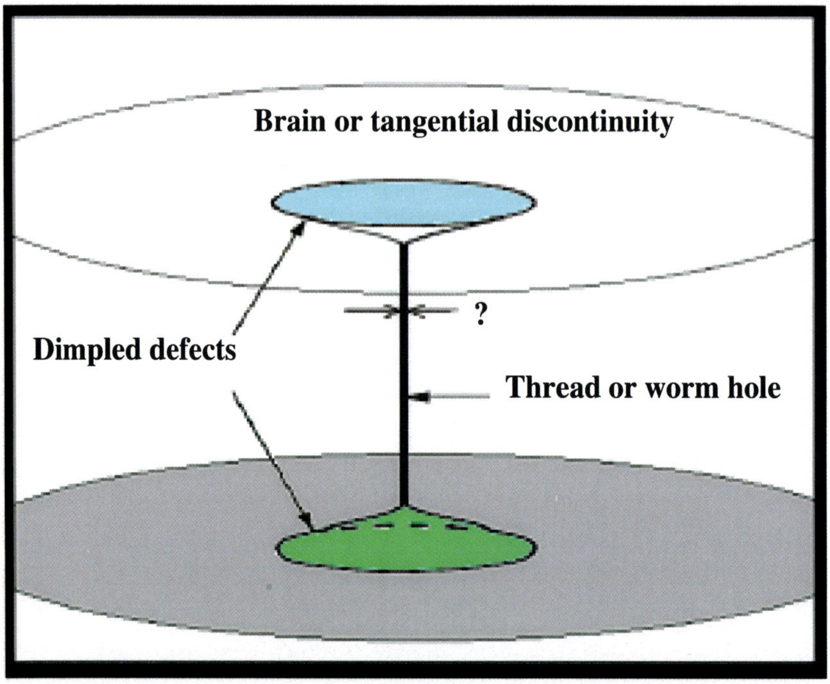

Figure 14. Falaco Solitons as connected dimples between Branes.

The new feature is that the "brane" surface of discontinuity is deformed by the Maximal surface dimple (which has been alluded to as a space-time foam [4]) This structure motivates the next section in which the idea is used to model the photon. The idea is also related to the rotational structures of rotating Bose-Einstein condensates [27].

In the next section, it is assumed that a thermodynamic (electromagnetic) system can be encoded by a 1-form of Action potentials, A, which leads by exterior differentiation to a 2-form of field intensities, $F = dA$. The null eigenvectors of the antisymmetric matrix representation of F will form 3D expanding spherical surfaces of propagating field discontinuities (related to the spatial portions of the Minkowski lightcone where $F\hat{}F = 0$). In addition, the isotropic Spinor eigenvectors of F will form surfaces of zero mean curvature as defect structures on the spherical spatial portions of the lightcone. The result is a Falaco Soliton pair (with $A\hat{}F \neq 0$) between the two bounding cycles of a spherical shell. The claim is that this concept serves as a model for the Photon.

5 Falaco Solitons As a Topological Model for a Photon

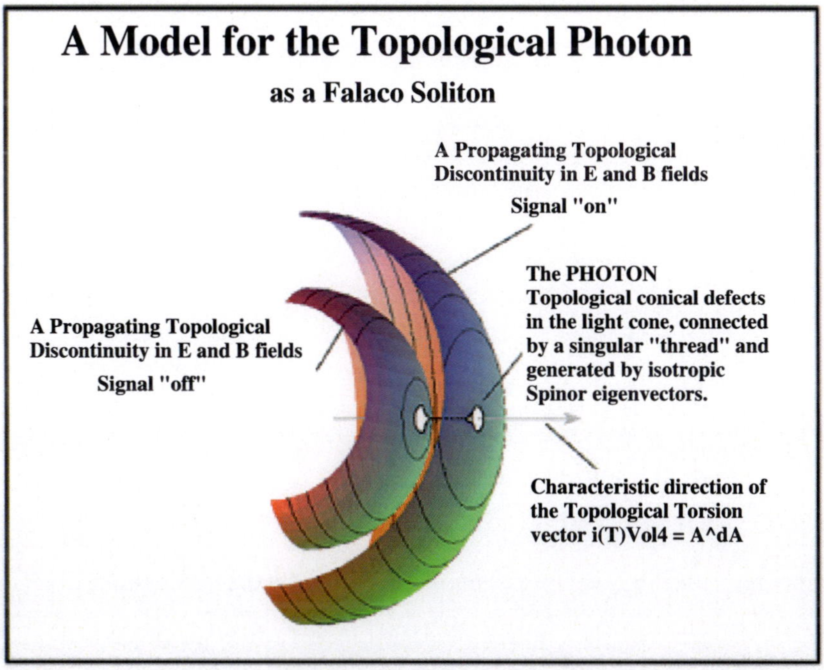

Fig. 15. The Photon as a Falaco Soliton between lightcone shells

The idea is to combine the topological features of the Minkowski signature, the possibilities of coherent states of "stationary" topology (solitons) for non-equilibrium, but thermodynamically closed, systems of Pfaff topological dimension 3 (with $A\hat{}F \neq 0$), and the fact that for such systems the electromagnetic 2-form, $F = dA$, has one pair of eigenvectors of eigenvalue zero, and one pair of complex conjugate isotropic null eigenvector arrays with imaginary eigenvalues. The eigenvectors with zero eigenvalues form Minkowski lightcones with $F\hat{}F = 0$. Consider two causal expanding spheres (two light cones) representing the "on" and "off" propagating discontinuity defects (as expanding concentric spheres in 3D). The concentric spherical surfaces of field discontinuity bound an interior region of finite electromagnetic field intensities, $\mathbf{E}$ and $\mathbf{B}$. The conjugate pair of Spinor eigenvectors define 2D surfaces of zero mean curvature as conical topological deformation defects on the light cones.

The region interior to the two light cone "shells" is a region of non equilibrium thermodynamics, where the Topological Torsion, $A\hat{}F$, is not zero, but has zero divergence. Hence the closed integrals of the three form, $A\hat{}F$, have rational ratios. That is, they are topologically quantized. It is suggested that these "quantized" torsion concepts are related to the concept of quantized orbital angular momentum introduced in recent investigations of photon systems. (The intrinsic spin is related to the 3-form $A\hat{}G$. [31]) The conical defects on each light cone are connected by a 1D "string", or "vortex tube", of zero radius, determined by the condition that evolution, V, in the direction of the components of the 3-form, $A\hat{}F$, of topological torsion, are extremal. That is, the thermodynamic work vanishes: $W = i(V)dA = i(V)F \Rightarrow 0$.

As an example, consider the 1-form of Action given by,

$$A = (m_e/e)\{\omega(xdy - ydx) - c^2 dt\}, \tag{72}$$

where the constants $(m_{electron}/e = h/(ec\lambda_{Compton})), \omega$ and c have been chosen on grounds of dimensional analysis. The Pfaff sequence demonstrates the the Pfaff topological dimension relative to the 1-form A is 3:

$$F = dA = (m_e/e)2\omega\, dx\hat{}dy = B_z dx\hat{}dy, \tag{73}$$

$$A\hat{}F = (m_e c^2/e)B_z dx\hat{}dy\hat{}dt, \tag{74}$$

$$F\hat{}F = 0. \tag{75}$$

There is no $\mathbf{E}$ field, but there is a $\mathbf{B}$ field component along the z axis. Hence the example has the properties

$$\mathbf{E} = 0,\ \mathbf{B} \neq 0,\ \mathbf{A} \neq 0,\ \ \mathbf{E} \circ \mathbf{B} = 0,\ \ \mathbf{A} \circ \mathbf{B} = 0 \tag{76}$$

Note that the scalar and vector potentials are given by the expressions

$$\phi = (m_e c^2/e), \tag{77}$$

$$\mathbf{A} = (m_e/e)\omega[-y, x, 0]. \tag{78}$$

The vector potential is tangent to a circle about the origin in the z = 0 plane. The direction field generated by the topological torsion vector is

$$\mathbf{T}_4 = [\mathbf{E} \times \mathbf{A} + \mathbf{B}\phi,\ \mathbf{A} \circ \mathbf{B}] \tag{79}$$

$$= [0, 0, B_z(m_e c^2/e), 0]. \tag{80}$$

For evolutionary processes $\mathbf{V}_4$ in the direction of $\mathbf{T}_4$, it follows that the Work 1-form vanishes.

$$W = i(\mathbf{V}_4)F = -\{\mathbf{E} + \mathbf{V} \times \mathbf{B}\} \circ d\mathbf{r} + (\mathbf{E} \circ \mathbf{V})dt, \quad (81)$$
$$W = i(\mathbf{T}_4)F = -\{\mathbf{0} + \mathbf{B}\phi\times\mathbf{B}\} \circ d\mathbf{r}+(0)dt \Rightarrow 0. \quad (82)$$

The evolutionary velocity field $\mathbf{V}$ in the direction of $\mathbf{T}_4$ is proportional to the $\mathbf{B}$ field.

This result gains credence from the observations of similar topological defects in fluid systems, the Falaco Solitons. Thermodynamic systems of Pfaff topological dimension 3 (based on the 1-form, A) are non equilibrium, thermodynamically closed systems that can exchange energy (radiation) but not mass with their environments . When the Photon is "created" the Pfaff topological dimension is presumed to be 4, with evolution along a space time direction field given by the Topological Torsion vector, $\mathbf{T}_4$. The processes is thermodynamically irreversible, and $(\mathbf{E} \circ \mathbf{B}) \neq 0$. The process evolves continuously to domains of Pfaff topological dimension 3, forming the "condensations" - or Photons - of topological coherence as stationary, but excited, states of a Hamiltonian process. It is conjectured that the conical topological defects are not constrained by a limiting speed C, but can move (transversely on the light cone) with speeds given by the projective Moebius transformations, which can vary from zero to infinity.

6 Continuous Topological Evolution

The objective of this section is to examine topological aspects and defects of thermodynamic physical systems and their possible continuous topological evolution, creation, and destruction on a cosmological scale. The creation and evolution of stars and galaxies will be interpreted herein in terms of the creation of topological defects of Pfaff topological dimension 3 and evolutionary phase changes in a very dilute turbulent, non-equilibrium, thermodynamic system of maximal Pfaff topological dimension 4. The cosmology so constructed is opposite in viewpoint to those efforts which hope to describe the universe in terms of properties inherent in the quantum world of Bose-Einstein condensates, super conductors, and superfluids [27]. Both approaches utilize the ideas of topological defects, but thermodynamically the approaches are opposite in the sense that the quantum method involves, essentially, equilibrium systems, while the approach presented herein is based upon non-equilibrium systems. Based upon the single assumption that the universe is a non-equilibrium thermodynamic system of Pfaff topological dimension 4 leads to a cosmology where the universe, at present, can be approximated in terms of the non-equilibrium states of a very dilute van der Waals gas near its critical point. The stars and the galaxies are the topological defects and coherent - but not equilibrium - structures of Pfaff topological dimension 3 in this non-equilibrium system of Pfaff topological dimension 4. The topological theory of the ubiquitous van der Waals gas leads to the concepts of negative pressure, string tension, and a Higgs potential as natural consequences of a topological point of view applied to thermodynamics. Perhaps of more importance is the fact that these concepts do not depend explicitly upon the geometric constraints of metric or connection, and yield a different perspective on the concept of gravity.

The original motivation for this conjecture is based on the classical theory of correlations of fluctuations presented in the Landau-Lifshitz volume on statistical mechanics [7].

However, the methods used herein are not statistical, not quantum mechanical, and instead are based on Cartan's methods of exterior differential forms and their application to the topology of thermodynamic systems and their continuous topological evolution [17]. Landau and Lifshitz emphasized that real thermodynamic substances, near the thermodynamic critical point, exhibit extraordinary large fluctuations of density and entropy. In fact, these authors demonstrate that for an almost perfect gas near the critical point, the correlations of the fluctuations can be interpreted as a 1/r potential giving a $1/r^2$ force law of attraction. Hence, as a cosmological model, the almost perfect gas - such as a very dilute van der Waals gas - near the critical point yields a reason for both the granularity of the night sky and for the $1/r^2$ force law ascribed to gravitational forces between for massive aggregates.

A topological (and non statistical) thermodynamic approach can be used to demonstrate how a four dimensional variety can support a turbulent, non-equilibrium, physical system with universal properties that are homeomorphic (deformable) to a van der Waals gas [28]. The method leads to the necessary conditions required for the existence, creation or destruction of topological defect structures in such a non-equilibrium system. For those physical systems that admit description in terms of an exterior differential 1-form of Action potentials of maximal rank, a Jacobian matrix can be generated in terms of the partial derivatives of the coefficient functions that define the 1-form of Action. When expressed in terms of intrinsic variables, known as the similarity invariants, the Cayley-Hamilton 4 dimensional characteristic polynomial of the Jacobian matrix generates a universal phase equation. Certain topological defect structures can be put into correspondence with constraints placed upon those (curvature) similarity invariants generated by the Cayley-Hamilton 4 dimensional characteristic polynomial. These constraints define equivalence classes of topological properties.

The characteristic polynomial, or Phase function, can be viewed as representing a family of implicit hypersurfaces. The hypersurface has an envelope which, when constrained to a minimal hypersurface, is related to a swallowtail bifurcation set. The swallowtail defect structure is homeomorphic to the Gibbs surface of a van der Waals gas. Another possible defect structure corresponds to the implicit hypersurface surface defined by a zero determinant condition imposed upon the Jacobian matrix. On 4 dimensional variety (space-time) , this non-degenerate hypersurface constraint leads to a cubic polynomial that always can be put into correspondence with a set of non-equilibrium thermodynamic states whose kernel is a van der Waals gas. Hence this universal topological method for creating a low density turbulent non-equilibrium media leads to the setting examined statistically by Landau and Lifshitz in terms of classical fluctuations about the critical point.

The conjecture presented herein is that non-equilibrium topological defects in a non-equilibrium 4 dimensional medium represent the stars and galaxies, which are gravitationally attracted singularities (correlations of fluctuations of density fluctuations) of a real gas near its critical point. Note that the Cartan methods do not impose (*a priori.)* a constraint of a metric, connection, or gauge, but do utilize the topological properties associated with constraints placed on the similarity invariants of the universal phase function.

Based upon the single assumption that the universe is a non-equilibrium thermodynamic system of Pfaff topological dimension 4 leads to a cosmology where the universe, at present, can be approximated in terms of the non-equilibrium states of a very dilute van der Waals gas near its critical point. The stars and the galaxies are the topological defects and coherent (but not equilibrium) self-organizing structures of Pfaff topological dimension 3 formed by irreversible topological evolution in this non-equilibrium system of Pfaff topological dimension 4.

The turbulent non-equilibrium thermodynamic cosmology of a real gas near its critical point yields an explanation for:

1. The granularity of the night sky as exhibited by stars and galaxies.
2. The Newtonian law of gravitational attraction proportional to $1/r^2$.
3. The expansion of the universe (4th order curvature effects).
4. The possibility of domains of negative pressure (explaining what has recently been called dark energy) due to a classical Higgs mechanism for aggregates below the critical temperature (3rd order curvature effects)
5. The possibility of domains where gravitational effects (2nd order Gauss curvature effects) appear to be related to entropy and temperature properties of the thermodynamic system.
6. The possibility of cohesion properties (explaining what has recently been called dark matter) due to string or surface tension (1st order Mean curvature effects)
7. Black Holes (generated by Petrov Type D solutions in gravitational theory [3]) are to be related to Minimal Surface solutions to the Universal thermodynamic 4th order Phase function.

7 Summary

As the Falaco phenomenon appears to be the result of a topological defect, it follows that as a topological property of hydrodynamic evolution, it could appear in any density discontinuity, at any scale. This rotational pairing mechanism, as a topological phenomenon, is independent from size and shape, and could occur at both the microscopic and the cosmic scales. In fact, as mentioned above, during the formative stages of the Falaco Soliton pair, the decaying Rankine vortices exhibit spiral arms easily visible as caustics emanating from the boundary of each vortex core. The observation is so striking that it leads to the conjecture: Can the nucleus of M31 be connected to the nucleus of our Milky way galaxy by a tubular cosmic thread? Can material be ejected from one galaxy to another along this comic thread? Can barred spirals be Spiral Arm galaxies at an early stage of formation - the bar being and exhibition of material circulating about the stabilizing thread? At smaller scales, the concept also permits the development of another mechanism for producing spin-pairing of electrons in the discontinuity of the Fermi surface, or in two dimensional charge distributions. Could this spin pairing mechanism, depending on transverse wave, not longitudinal wave, coupling be another mechanism for explaining superconductivity? As the defect is inherently 3-dimensional, it must be associated with a 3-form of Topological Torsion, A^dA, introduced by the author in 1976 [13] [15] [16] [20], but now more commonly called a Chern Simons term. These ideas were exploited in an attempt to explain high TC superconductivity [18]. To this author the importance of the Falaco Solitons is that they offer the first clean experimental evidence of topological defects taking place in a dynamical system. Moreover, the experiments are fascinating, easily replicated by anyone with access to a swimming pool, and stimulate thinking in almost everyone that observes them, no matter what his field of expertise. They certainly are among the most easily produced solitons.

More detail (with downloadable pdf files of almost all publications) may be found on the web site:

http://www.cartan.pair.com

The original observation was first described at a Dynamics Days conference in Austin, TX, [14] and has been reported, as parts of other research, in various hydrodynamic publications, but it is apparent that these concepts have not penetrated into other areas of research. As the phenomenon is a topological issue, and can happen at all scales, the Falaco Soliton should be a natural artifact of both the sub-atomic and the cosmological worlds. The reason d'etre for this short article is to bring this idea to the attention of other researchers who might find the concept interesting and stimulating to their own research

References

[1] Bryant, R.L.,Chern, S.S., Gardner, R.B.,Goldschmidt, H.L., and Griffiths, P. A. (1991) *"Exterior Differential Systems"*, Springer, Verlag.

[2] Cartan, E. (1966) *"The Theory of Spinors"*, Dover, New York, p. 39.

[3] Chandrasekar,.S. *"The mathematical theory of Black Holes"*, (Clarendon press, Oxford, N.Y. 1983)

[4] Dzhunushaliev, V., (2002) "*Strings in the Einstein's paradigm of matter*", http://arxiv.org/abs/gr-qc/0205055

[5] Fernandez, I., Lopez, F. J., Souam, R. *"The space of complete embedded maximal surfaces with isolated singularities in the 3-dimensional Lorentz-Minkowski space L^3"*, *arXiv:math.DG/0311330* Rev v2 27 JAN 2005

[6] Landau, L. D. and Liftshitz, E. M. (1959) *"Fluid Mechanics"*, Pergamon p. 94.

[7] Landau, L. D. and Lifshitz, E. M., (1958), *"Statistical Physics"* Pergamon Press, London p. 369.

[8] Landau, L. D. and Lifshitz, E. M. (1960), *"Electrodynamics of Continuous Media"*, Pergamon Press, London, p. 97.

[9] Langford, W. F. (1983) in *"Non linear Dynamics and Turbulence"* Edited by G.I. Barrenblatt, et. al. Pitman, London

[10] http://www22.pair.com/csdc/download/maxlor.pdf

[11] Pedoe. D., (1988) *"Geometry, A comprehensive course"*, Dover, NY p.276

[12] Osserman, R. (1986) *"A Survey of Minimal Surfaces"*, Dover, N.Y. 1986)

[13] Kiehn, R.M. (1976), *NASA AMES NCA-2-OR-295-502* (http://www22.pair.com/csdc/pdf/rmkames.pdf)

[14] Kiehn, R.M, (1987) "The Falaco Effect, A Topological Soliton" Talk presented at the Dynamics Days conference, Austin, TX Jan. 1987. *The easily reproduced experiment added to the credence of topological defects in fluids. It is now perceived that this topological phenomenon is universal, and will appear at all levels from the microscopic to the galactic.*

[15] Kiehn, R.M. (1990), *Topological Torsion, Pfaff Dimension and Coherent Structures*, in: *"Topological Fluid Mechanics"*, H. K. Moffatt and T. S. Tsinober eds, Cambridge University Press p. 449-458. (../camb89.pdf)

[16] Kiehn, R. M. (1991), *Compact Dissipative Flow Structures with Topological Coherence Embedded in Eulerian Environments*, in: *"Non linear Dynamics of Structures"*, edited by R.Z. Sagdeev, U. Frisch, F. Hussain, S. S. Moiseev and N. S. Erokhin, p.139-164, World Scientific Press, Singapore

[17] Kiehn, R.M. (1991) "*Continuous Topological Evolution*", *LANL arXiv/math-ph/0101032* (../contevol3.pdf)

[18] Kiehn, R. M. (1991), *Are there three kinds of superconductivity*, *Int. Journ. of Modern Physics*, vol5 #10 p. 1779

[19] Kiehn, R.M. (1991) *Some Closed Form Solutions to the Navier Stokes Equations*, *LANL arXiv/physics/0102002* (../nvsol.pdf)

[20] Kiehn, R. M. (1992), *Topological Defects, Coherent Structures and Turbulence in Terms of Cartan's Theory of Differential Topology*, in *"Developments in Theoretical and Applied Mathematics, Proceedings of the SECTAM XVI conference"*, B. N. Antar, R. Engels, A.A. Prinaris and T. H. Moulden, Editors, The University of Tennessee Space Institute, Tullahoma, TN 37388 USA.

[21] Kiehn, R. M. (1993), *Instability Patterns, Wakes and Topological Limit Sets*, in "*Eddy Structure Identification in Free Turbulent Shear Flows*", J.P.Bonnet and M.N. Glauser, (eds), Kluwer Academic Publishers p. 363.

[22] Kiehn, R. M., (1999), *Coherent Structures in Fluids are Topological Torsion Defects*, in *J*, "*IUTAM Symposium on Simulation and Identification of Organized Structures in Flows*",.N. Sørensen, et al., eds., Kluwer Academic Publishers, Dordrecht,. See (../copen5.pdf). Presented at the IUTAM-SIMFLO Conference at DTU, Denmark, May 25-29, (1997).

[23] http://www22.pair.com/csdc/download/spotsmovie.avi

[24] Struik, D. (1961) "Differential Geometry", Addison Wesley, Reading, Mass

[25] Tisza, L. (1966) "*Generalized Thermodynamics*", MIT press, Cambridge, p.125.

[26] Tornkvist, O. and Schroeder, E., (1997) *Vortex dynamics in Dissipative Systems*, *Phys. Rev. Lett.* **78**, 10, p. 1908-1911.

[27] Volovik, G. E., (2003) *"The Universe in a Helium droplet"*, Claredon Press, Oxford, 2003

[28] Kiehn, R. M. (2004) *Non Equilibrium Thermodynamics*, *"Non Equilibrium Systems and Irreversible Processes Vol 1"* see http://www.cartan.pair.com

[29] Kiehn, R. M. (2004) *Cosmology, Falaco Solitons and the Arrow of Time*, *"Non Equilibrium Systems and Irreversible Processes Vol 2"* see http://www.cartan.pair.com

[30] Kiehn, R. M. (2004) *Wakes, Coherent Structures and Turbulence*, *"Non Equilibrium Systems and Irreversible Processes Vol 3"* see http://www.cartan.pair.com

[31] Kiehn, R. M. (2004) *Plasmas and Non equilibrium Electrodynamics*, *"Non Equilibrium Systems and Irreversible Processes Vol 4"* see http://www.cartan.pair.com

[32] Kiehn, R. M. (2004) *Exterior Differential Forms and Differential Topology*, *"Non Equilibrium Systems and Irreversible Processes Vol 5"* see http://www.cartan.pair.com

[33] Kiehn, R. M. (2004) *Maple Magic and Differential Topology* *"Non Equilibrium Systems and Irreversible Processes Vol 6"* see http://www.cartan.pair.com

[34] Zeldovich, Ya B. and Novikov, I. D. (1996) *"Stars and Relativity"*, Dover, New York

INDEX

A

B

C

D

E

F

G

H

I

J

K

L

M

N

T

U

V

W

Y